What's the Matter with Waves?

An introduction to techniques and applications of quantum mechanics

Series on wave phenomena in the physical sciences

Series Editor

Sanichiro Yoshida

Southeastern Louisiana University

About the series

The aim of this series is to discuss the science of various waves. It consists of several books, each covering a specific subject known as a wave phenomenon. Each book is designed to be self-contained so that the reader can understand the gist of the subject. From this viewpoint, the reader can read any book as a stand-alone article. However, it is beneficial to read multiple books as it would provide the reader with the opportunity to view the same aspect of wave dynamics from different angles.

The targeted readership is graduate students of the field and engineers whose background is similar but different from the subject. Throughout the series, it is intended to help students and engineers deepen their fundamental understanding of the subject as wave dynamics. An emphasis is laid on grasping the big picture of each subject without dealing with detailed formalism, and yet understanding the practical aspects of the subject. To this end, mathematical formulations are simplified as much as possible and applications to cutting edge research are included. The reader is encouraged to read books cited in each book for further details of the subject.

Other titles in this series

Sanichiro Yoshida *Waves: Fundamentals and Dynamics*

Wayne D Kimura *Electromagnetic Waves and Lasers*

David Feldbaum *Gravitational Waves*

Michail Todorov *Nonlinear Waves: Theory, Computer Simulation, Experiment*

What's the Matter with Waves?

An introduction to techniques and applications of quantum mechanics

William Parkinson

Southeastern Louisiana University

Morgan & Claypool Publishers

Rights & Permissions
To obtain permission to re-use copyrighted material from Morgan & Claypool Publishers, please contact info@morganclaypool.com.

ISBN   978-1-6817-4577-0 (ebook)
ISBN   978-1-6817-4578-7 (print)
ISBN   978-1-6817-4579-4 (mobi)

DOI   10.1088/978-1-6817-4577-0

Version: 20171201

IOP Concise Physics
ISSN 2053-2571 (online)
ISSN 2054-7307 (print)

A Morgan & Claypool publication as part of IOP Concise Physics
Published by Morgan & Claypool Publishers, 1210 Fifth Avenue, Suite 250, San Rafael, CA, 94901, USA

IOP Publishing, Temple Circus, Temple Way, Bristol BS1 6HG, UK

This book is dedicated to the most important people in my life—my family, especially my wife Bonnie.

Contents

Author biography

Bill Parkinson

Bill Parkinson attended California University of Pennsylvania, receiving a BS in chemistry in 1977. This was followed with stints as an environmental engineer, a construction worker, marine biologist, and high school physics and mathematics teacher. He obtained his PhD from the University of Florida's Quantum Theory Project in 1989, where he had the great fortune of rubbing elbows with the world's leading experts in computational chemistry during some of the field's most formative years. After postdoctoral positions at Odense University (now Syddansk Universitet, the University of Southern Denmark) and Texas A&M, he joined the faculty of Southeastern Louisiana University in 1991. His pastimes and passions include yard work, biking, volleyball, the beach, and Pittsburgh Steeler football.

What's the Matter with Waves?
An introduction to techniques and applications of quantum mechanics
William Parkinson

Chapter 1

Introduction

Everyday experience is a helpful guide as we attempt to model the physical world around us. For example, interpreting wave motion is aided by noting the movement of a bobber on the ripples of a pond. Anticipating trajectories resulting from colliding bodies may be facilitated from observation of a pool table cue ball. Understanding action at a distance from a force field is furthered by participating in a game of catch. With regard to the subject at hand, observing ocean water crashing into a rock jetty, listening to sound echoing through a mountain valley, or locating a penny on the bottom of a fountain provides a basis of comprehension for wave–matter interaction.

By contrast attempts to phenomenologically model a chemical process in which matter undergoes a transformation of constitution present daunting challenges. Ignoring for a moment the fact that reaction rates commonly occur in a time frame that render them humanly imperceptible, a more fundamental impediment is the scale of matter dimensionally. We cannot 'see' the system interacting, we may only verify the occurrence of chemistry from observing affects on its surroundings: heat evolution, a phase change, a flash of light, a smell, a puff of smoke. In fact, the only 'illuminating' probe at these dimensions is a light wave. Suppose then that we are somehow magically able to ride a light beam, as if it allows us to don nanoscopes for the purpose of observing matter in action during a chemical process. Even with this advantage, we would quickly find that attempts to utilize rules for cue balls or other projectile motions do not apply. We would also learn the light itself is not just a casual observer in this environment, but is an intimate part of the system dynamics.

Ultimately would come to the realization that an entirely new set of principles and guidelines, far outside the box of those ingrained from familiar observation, are required to correctly model and predict events. The approach taken here will then be to review our understanding of the behavior of macroscopic matter. Particular focus will be given to instances where the governing rules hold fast, and where there is a disconnect. This is facilitated by reviewing experimental evidence that could not be

doi:10.1088/978-1-6817-4577-0ch1

explained away by accepted guidelines. We will then introduce the necessary modifications to allow matter's description at the atomic and molecular level.

While navigating this course, the advantage taken from tangible connection with everyday experience must unfortunately be abandoned. We will in fact find it necessary to incorporate some facets of our common understanding of wave behavior into the model of matter. This seems counterintuitive. Even to the most casual observer, there are obvious differences between matter and waves. Though each takes a variety of distinguishable forms, matter and waves are ultimately differentiated by a single criterion. Waves uniquely have the capability of occupying the same space at the same time. For example consider four individuals simultaneously conducting two separate conversations as depicted in figure 1.1. Everyday experience tells us that opposing pairs can communicate, even though sounds from their voices are somewhere intersecting. In addition, light reflecting off any one of them can be detected by the remaining three, even though these waves must also inhabit the same space along their journey. This property, known as *superposition*, allows waves to exhibit constructive and destructive interference, which for sound results in phenomena such as piano chords and devices like noise-cancelling headphones.

By distinct contrast, one of the two fundamental characteristics of matter is that it 'takes up space' (the other, according to any physical science primer, being that it has mass. This is technically saying the same thing actually, but I digress...). Of course the implication of matter occupying space is that it must exclude other matter from that space. Despite this very fundamental of differences, matter and waves are intimately related. Most waves are actually disturbances of matter. Sound, water, and string waves cannot move through space without matter acting as a *medium* to

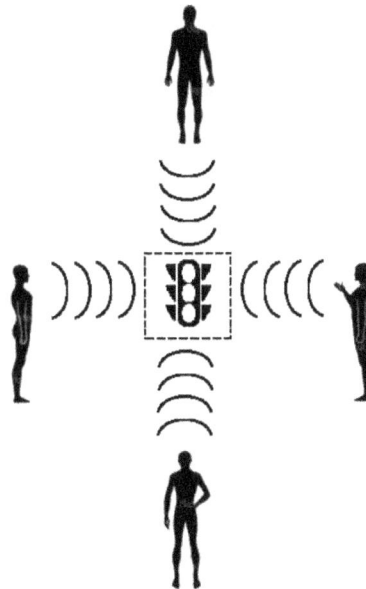

Figure 1.1. Superposition of propagating waves.

enable their propagation. In fact for quite some time it was believed waves and matter had an inextricable relationship. With Huygens' introduction of the wave theory of light, it presupposed the requirement of a support medium. The search ensued for the so-called *aether*, until its existence was debunked as a result of experiments performed by Michelson and Morley.

In contrast to other wave forms, electromagnetic radiation creates a disturbance in *space*. This effect is described in physics as being generated by a *field*, or interacting fields as is the case here. In fact, electromagnetic radiation prefers an absence of matter. As evidenced by gazing upward towards the nighttime sky, electromagnetic radiation continuously self-generates and propagates through space when unimpeded by matter. When electromagnetic radiation of a particular type does encounter matter it may pass through, be diffracted, refracted, reflected, scattered, or absorbed. Matter and its interaction with light and other portions of the electromagnetic spectrum is the basis of a wide variety of qualitative and quantitative analytical chemistry techniques.

The arrival of spectroscopy as a characterizing tool for the composition of substances was paralleled by the fundamental questions light–matter interaction posed. By the late nineteenth and early twentieth centuries, instrumentation had achieved levels of resolution revealing information that could not be explained by contemporary theories. Attempts to formalize the mechanism of light–matter energy transfer blurred their fundamental distinction. As an initial explanation, light waves were re-imagined to possess matter-like characteristics. Eventually viewpoints shifted to treating matter from a wave perspective. In this way many experimental inconsistencies could be resolved. Ultimately, we should be resigned to the fact that both light or matter can individually exhibit wave- or particle-like characteristics. The circumstances dictating their behavior ultimately depend on the situation, but invariably occur at the atomic and molecular level. Our purpose in subsequent sections is therefore to take a nanoscopic view of matter behaving as a wave in order to gain insight into its macroscopic properties.

As a prelude to the tale of matter and waves and its timeline, we must begin two centuries beforehand, to properly acknowledge antecedent milestones. The eighteenth century is appropriately known as the 'Age of Enlightenment' or 'Age of Reason.' With apology to the scientific and engineering accomplishments of the period, an incredible array of mathematical techniques and advancements were introduced by Gauss, Euler, Fourier, LaPlace, Maclaurin, LaGrange, Taylor, Leibnitz, Bernoulli, Legendre, Newton and others. At the time much of this probably seemed no more than academic indulgence with little or no connection to the real world. However a century later these techniques were essential to formulation of thermodynamics and electrodynamics by individuals including Ampere, Faraday, Maxwell, Clausius, Joule, Helmholtz, Boltzmann, Thompson (Lord Kelvin), and Gibbs.

These achievements marked a seminal moment in the annals of scientific accomplishment. After a lengthy gestation, a coming of age was signaled. With roots in human curiosity, fear, and superstition followed by a lengthy infancy of straightforward phenomenological modeling, science now embraced a new, fundamental purpose. Interpreting the workings of the everyday world was no longer the be-all and end-all. Scientists pushed the envelope of human experience to dimensions beyond what could be seen by a telescope or microscope. The frontiers of science were inextricably dependent on abstract mathematical techniques, culminating in more versatile, robust, and predictive scientific models. Theory now blazed a trail for experimental investigation.

Throughout this time, both the interpretation of matter as well as the properties and behavior of waves were thought to be on firm theoretical ground. However, evidence which emerged in the late nineteenth and early twentieth centuries blurred the lines between matter and waves. After much consternation, reflection, and debate among scientists, a blended behavior of light and matter emerged known as *wave–particle duality*. Much of the same mathematics, that by then had served theoreticians so well for over a century, again proved crucial and indispensable.

Quantum mechanics, the consummation of wave–matter interaction, marked a paradigm shift in more ways than the radical departure of its physics. It did not congeal from the singular ruminations or epiphany of any one individual. Quantum mechanics was an evolution of thought and philosophy coalesced from decades of work, culminating from efforts of an unprecedentedly-large collection of vital contributors. Previous landmark events in science could almost invariably be attributed to efforts of a single individual. The global effort that quantum mechanics represented was a testament to evolving human connectivity. By the twentieth century, scientists were taking full advantage of information sharing from advances in communication and experiencing an increased mobility attributable to ease of travel.

Quantum mechanics refined our understanding of matter to the point that its profound impact now demarcates the advent of 'modern physics.' Within a brief period of time it reverberated across chemistry and molecular biology as well. Subsequent to its introduction several of the leading scientists of the day, most familiarly Einstein, took on worldwide celebrity status. Nobel prizes were awarded to a variety of its principal contributors over a broad span of the twentieth century. These include prizes in physics to Planck in 1918, Einstein in 1921, Bohr in 1922, de Broglie in 1929, Heisenberg in 1932, Schrödinger in 1933, Pauli in 1945, and Born in 1954. Awards for contributions of quantum mechanics in chemistry were given to Pauling in 1954, Mullikan in 1966, and Pople and Kohn in 1998. Many other recipients in both fields were either guided in their experiments or directly impacted in their theoretical developments by quantum mechanics. It is somewhat unsettling to read the press release accompanying Mullikan's 1966 prize which points out the overwhelmingly complex nature of the discipline, essentially stating that quantum mechanics was inaccessible to the layperson. One of the main goals of this work is to help allay such predispositions or trepidations.

To punctuate the human interest aspect, no other image heralds the arrival of quantum mechanics or underscores the collective effort behind it quite like figure 1.2, a photograph of participants in the 1927 Solvay Conference. These invitation-only events feature varying focus topics that to this day they are intermittently held in Brussels, having been instituted by Belgian industrialist Ernest Solvay in 1911. The

Figure 1.2. Participants of the 1927 Solvay Conference.
Row 1: I Langmuir, M Planck, M Sklodowska-Curie, H Lorentz, A Einstein, P Langevin, C Guye, C Wilson, C Richardson
Row 2: P Debye, M Knudsen, W Bragg, H Kramers, P Dirac, A Compton, L de Broglie, M Born, N Bohr
Row 3: A Piccard, E Henriot, P Ehrenfest, E Herzen, T de Donder, E Schrödinger, J Verschaffelt, W Pauli, W Heisenberg, R Fowler, L Brillouin

1927 meeting, fifth in the series, featured lectures and discussions focused on the title subject: 'Electrons and Photons,' and a conference theme parallel to the topics of this book. Lewis once wrote, 'Science has its cathedrals, built by the efforts of few architects and of many workers.' The 1927 Solvay Conference validated one's standing as an architect to the sanctum of quantum mechanics. Essentially everyone who was anyone relevant to its development was present, a contingent in some ways analogous to the 1992 US men's Olympic basketball 'Dream Team.' Even those with no more than a passing knowledge of science will recognize several names. Those with a passion for it should particularly appreciate the special nature of the moment.

IOP Concise Physics

What's the Matter with Waves?
An introduction to techniques and applications of quantum mechanics
William Parkinson

Chapter 2

Motion in matter

We begin the study with a discussion of classical mechanics, developed using the approach introduced by Hamilton. Similar to Lagrangian mechanics, it is a re-formulation of the traditional Newtonian approach. In certain physical situations these alternatives provide insight that Newton's kinematics lack. Hamilton's formulation is of particular utility, as it lends itself seamlessly as we transition to quantum mechanics. The central feature of this approach is the Hamiltonian H. This function contains information describing the energy content of a particle or system of particles for all times. As is the case in most physical situations, the Hamilton can be partitioned into kinetic (T) and potential (V) energy components such that:

$$H = T + V \tag{2.1}$$

Time evolution of a system's ith particle is given by Hamilton's equations:

$$\frac{dp_i}{dt} = -\frac{\partial H}{\partial q_i} \tag{2.2}$$

$$\frac{dq_i}{dt} = \frac{\partial H}{\partial p_i} \tag{2.3}$$

In equations (2.2) and (2.3), p and q are the momentum and generalized position coordinates, respectively.

We shall begin by considering two objects of mass m_1 and m_2 in motion through space with trajectory vectors $\vec{r}_1$ and $\vec{r}_2$, respectively (see figure 2.1). We make the assumption that the particles experience no external potential ($V = 0$). With momentum defined as $p = mv$, the scalar kinetic energy of each particle is:

$$T_i = \frac{|\vec{p}_i|^2}{2m_i} \tag{2.4}$$

doi:10.1088/978-1-6817-4577-0ch2 2-1

Hamilton's equations are consistent with the system's kinematics. According to equation (2.3), the velocity of each particle is:

$$v_i = \frac{\mathrm{d}r_i}{\mathrm{d}t} = \frac{\partial H}{\partial p} = \frac{p_i}{m_i} \tag{2.5}$$

This two body system has six *degrees of freedom*. Either particle can independently translate in any of three Cartesian directions of motion: x, y, or z or, from the perspective of spherical polar coordinate system, independently in components: r, θ, or ϕ. We assume the motion of each mass is constrained by a binding force. Referring to figure 2.1, three degrees of freedom describe the unit vector components of $\vec{R}$ for translation of the center of mass M ($= m_1 + m_2$). The magnitude of $\vec{R}$ is determined from the mass-weighted averages of $\vec{r_1}$ and $\vec{r_2}$:

$$\vec{R} = \frac{m_1 \vec{r_1} + m_2 \vec{r_2}}{m_1 + m_2} \tag{2.6}$$

The remaining three degrees of freedom express components of $\vec{r}$, the internal motion vector (a chemist would call its magnitude the 'bond length'). From figure 2.1, vector addition gives: $\vec{r_2} = \vec{r_1} + \vec{r}$. When substituted into equation (2.6) with rearrangement, $\vec{r_1}$ can be expressed in terms of $\vec{r}$ and $\vec{R}$:

$$\vec{r_1} = \vec{R} - \frac{m_2}{m_1 + m_2}\vec{r} \tag{2.7}$$

Similarly, from figure 2.1 we note that: $\vec{r_1} = \vec{r_2} - \vec{r}$. Substituting this into equation (2.6) gives:

$$\vec{r_2} = \vec{R} + \frac{m_1}{m_1 + m_2}\vec{r} \tag{2.8}$$

Equations (2.7) and (2.8) are substituted into the kinetic energy expression:

$$T = \frac{|\vec{p_1}|^2}{2m_1} + \frac{|\vec{p_2}|^2}{2m_2} = \frac{1}{2}\left[m_1 \left| \frac{\mathrm{d}\vec{r_1}}{\mathrm{d}t} \right|^2 + m_2 \left| \frac{\mathrm{d}\vec{r_2}}{\mathrm{d}t} \right|^2 \right] \tag{2.9}$$

Expanding the dot products leads to:

$$\begin{aligned} T = \frac{1}{2}m_1 &\left[\left| \frac{\mathrm{d}\vec{R}}{\mathrm{d}t} \right|^2 - 2\frac{m_2}{m_1 + m_2}\frac{\mathrm{d}\vec{R}}{\mathrm{d}t} \cdot \frac{\mathrm{d}\vec{r}}{\mathrm{d}t} + \frac{m_2^2}{(m_1 + m_2)^2}\left| \frac{\mathrm{d}\vec{r}}{\mathrm{d}t} \right|^2 \right] \\ + \frac{1}{2}m_2 &\left[\left| \frac{\mathrm{d}\vec{R}}{\mathrm{d}t} \right|^2 + 2\frac{m_1}{m_1 + m_2}\frac{\mathrm{d}\vec{R}}{\mathrm{d}t} \cdot \frac{\mathrm{d}\vec{r}}{\mathrm{d}t} + \frac{m_1^2}{(m_1 + m_2)^2}\left| \frac{\mathrm{d}\vec{r}}{\mathrm{d}t} \right|^2 \right] \end{aligned} \tag{2.10}$$

Equation (2.10) is simplified by noting the cross-term derivatives involving $\vec{r}$ and $\vec{R}$ cancel, and that:

$$\frac{1}{2}m_1\frac{m_2^2}{(m_1 + m_2)^2} + \frac{1}{2}m_2\frac{m_1^2}{(m_1 + m_2)^2} = \frac{1}{2}\frac{m_1 m_2}{m_1 + m_2} = \frac{1}{2}\mu \tag{2.11}$$

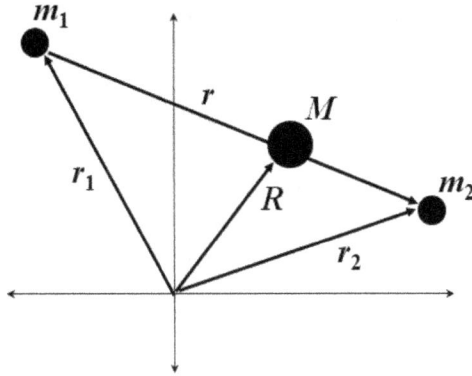

Figure 2.1. Coordinate system of a two-body problem.

Equation (2.11) introduces the *reduced mass* of the two-body system μ, which can alternatively be defined:

$$\frac{1}{\mu} = \frac{1}{m_1} + \frac{1}{m_2} \tag{2.12}$$

Note from either equation (2.11) or (2.12), as: $m_1 \to m_2$, $\mu \to \frac{1}{2}m_1$, and as $m_1 \to 0$, $\mu \to m_1$. Equation (2.10) expressed as a function of the center of mass M, the center of mass vector $\vec{R}$, the reduced mass μ, and the internal motion vector $\vec{r}$ is thus:

$$T = \frac{1}{2}M \left| \frac{\mathrm{d}\vec{R}}{\mathrm{d}t} \right|^2 + \frac{1}{2}\mu \left| \frac{\mathrm{d}\vec{r}}{\mathrm{d}t} \right|^2 \tag{2.13}$$

The interpretation of equation (2.13) is as follows. For a two-body system, three of the six possible degrees of freedom result from changes in the Cartesian components of $\vec{R}$. This is the kinetic energy due to translation of the center of total mass M (and inherently properties such as the kinetic temperature and pressure of a bulk sample of these systems). The remaining three degrees of freedom are contributions to T due to changes in the Cartesian directions of the internal motion vector $\vec{r}$. It is even more instructive when these degrees of freedom are connected to the rate of change of $\vec{r}$, which occur in either one of two ways. If only the *magnitude* of $\vec{r}$ changes with time, the internal motion vector is describing relative *vibrational* motion of the two bodies. Alternatively, as the *direction* of $\vec{r}$ changes, the internal motion vector is undergoing two-body *rotational* motion.

The six independent motions are now constrained to three assigned to translation of the center of mass, and the three that remain representing one vibrational mode (as $\vec{r}$ expands and contracts) and two degenerate rotational modes (as $\vec{r}$ changes direction centered on one of the system's two equivalent moments of inertia). The latter three are independent of the translational degrees of freedom as is evidenced by no change in the center of mass location during these internal motions. As a matter of fact, the center of mass must be invariant to any purely rotational or vibrational mode independent of the number of coupled particles in the system. However, rotation and vibration are not strictly independent of each other.

Vibration results in a changing moment of mass inertia, which to maintain conservation of angular momentum requires an accompanying change in angular rotation speed. Although this complication can be addressed by perturbation theory techniques, we currently hand-wave our way out of this dilemma by assuming a *rigid rotor* approximation, wherein the scalar component of $\vec{r}$ does not change as it changes direction (e.g. a rotating diatomic molecule with fixed bond length).

The reduced mass μ mathematically expresses the mechanics of a two-body system as a one-body system. This simplifies the perspective of vibrational motion from the change in distance between two masses m_1 and m_2 into the movement of a single mass of value μ relative to an infinitely-massive fixed point. Likewise for rotation, instead of thinking about the concerted rotational dance of m_1 with m_2, we can picture mass μ moving circularly relative to an infinitely-massive fixed point in space.

IOP Concise Physics

What's the Matter with Waves?
An introduction to techniques and applications of quantum mechanics
William Parkinson

Chapter 3

Vibrating matter

3.1 Classical vibration

For reasons more pedagogical than pedantic, we begin our discussion of coupled mass motion by considering vibration. In fact, it was during attempts to mathematically model experimentally observed electromagnetic wave generation by vibrating matter that irreparable flaws were exposed in the interpretation of wave matter interaction. In addition solutions to the differential equations of a classically-treated harmonic oscillator serve as a relatively straightforward introduction to many of the mathematical techniques that will be required as we transition to quantum mechanics.

The oscillator is constructed by attaching mass μ to an ideal spring, one which suffers no dissipative loss of energy, that in turn is connected to an infinitely-massive wall. As discussed in chapter 2, μ in this case is exactly m, the mass of the oscillator. Using μ reminds us that the oscillator can indeed be a two-body system. When stretched and released, μ executes 1-dimensional un-damped vibrational motion such that both the amplitude and rate of oscillation are constant with time. This constitutes a *harmonic oscillator* with a pattern of movement known as *simple harmonic motion*. As un-physical as this may seem, it is exhibited by the nuclei of molecules in a variety of modes. Harmonic motion is sustained by the driving force of ambient temperature, and can be stimulated to different oscillations as matter absorbs particular frequencies of electromagnetic radiation. Before we examine the nature of this motion in molecules, it is instructive to first treat a model harmonic oscillator *classically* within Hamilton's framework. Our efforts will not be wasted for two important reasons. First, the exercise will use many of the same mathematical approaches and defining terms that will arise in quantum mechanical systems. More importantly, we will see there is an inconsistency to the energy distribution in a classical oscillator compared to the energy of interaction between electromagnetic waves and vibrating matter.

The attached spring supplies *a restoring force* relative to the direction of motion of the mass, according to the empirical rule known as Hooke's Law:

$$\vec{F}_x = -k\vec{x} \tag{3.1}$$

The minus sign of equation (3.1) indicates the direction of the restoring force, hence the acceleration experienced by the object, is in opposition to the mass displacement. The scalar proportionality factor k is known as the *spring constant*, which by dimensional analysis must possess units: $N \cdot m^{-1}$ or the SI form: $kg \cdot s^{-2}$. The magnitude of k describes the spring 'stiffness'. Equation (3.1) demonstrates that a spring with large k requires a large applied force to achieve small displacement, while a spring with small k will exhibit substantial displacement with small applied force. In ballpark terms, a Slinky has spring constant of roughly $k = 0.5 \, N \cdot m^{-1}$, contrasted by a car shock absorber with $k \approx 50\ 000 \, N \cdot m^{-1}$. Bond strengths of molecules are inferred from their gas phase spring constants (referred to by chemists as bond force constants). Representative examples are provided by the relatively weakly-bound hydrogen chloride (HCl): $k = 480 \, N \cdot m^{-1}$ in contrast to carbon monoxide (CO): $k = 1860 \, N \cdot m^{-1}$.

The oscillator's kinetic energy is related to the momentum of mass μ. Its potential energy is dependent on the displacement of μ, and is symmetric independent of the type of displacement (whether the spring is stretched or compressed). The system Hamiltonian takes the following form:

$$H = T(p) + V(x) \tag{3.2}$$

From chapter 2, equation (2.2) relates the rate of change of momentum and potential:

$$\frac{dp}{dt} = -\frac{\partial H}{\partial x} = -\frac{\partial V(x)}{\partial x} \tag{3.3}$$

Using the momentum expression $p = mv$, and Newton's Second Law written in terms of rate of change of momentum, we have:

$$F = -\frac{\partial V(x)}{\partial x} \tag{3.4}$$

Equation (3.4) is true of any *conservative force* like an ideal spring, or a conservative field, such as those generated by gravity, point charges, or magnets. Equation (3.1) is inserted in equation (3.4), rearranged and integrated over definite limits to give:

$$V(x) = k \int_0^x x' \, dx' = \frac{1}{2}kx^2 \tag{3.5}$$

The spring stores no potential energy at zero displacement (the bottom limit of the left-hand side of equation (3.5) integrates to $V(0) = 0$), and at other values is parabolic (see figure 3.1). The 'steepness' of the parabolic potential is directly proportional to the spring constant k. Stiff springs exhibit sharply increasing potential energy with displacement while floppy springs have relatively shallow potential curves.

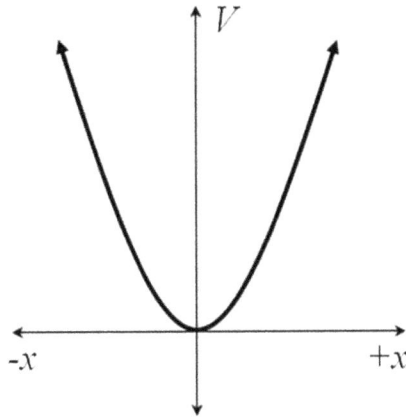

Figure 3.1. The potential energy curve of a harmonic oscillator.

An analytical expression for the displacement x is found by solving the differential equation that arises from applying Newton's Second Law:

$$F_x = \frac{dp_x}{dt} = \mu a_x = \mu\frac{dv_x}{dt} = \mu\frac{d^2x}{dt^2} \tag{3.6}$$

and equation (3.1):

$$\frac{d^2x}{dt^2} = -\frac{k}{\mu}x \tag{3.7}$$

Expressions in the form of equation (3.7), known as *eigenvalue problems*, play an important role in applied physics and mathematics in general. Variable x, *the eigenfunction*, is subjected to a mathematical action known as an *operator*, which in this case is the process of taking a second derivative: d^2/dt^2. In fact, this action can be viewed as sequential applications of a first derivative: $d^2/dt^2 = d/dt(d/dt)$. As seen in equation (3.7), the operator acts on an eigenfunction, and returns the eigenfunction along with its *eigenvalue*, in this case: $-k/\mu$. Eigenvalue equations are central to quantum mechanics, but as this example shows also have practical utility across physics and applied mathematics.

The solution to equation (3.7) requires an eigenfunction which upon taking its second derivative returns the negative of that eigenfunction. Both a real and imaginary general solution can be proposed:

$$x = \begin{matrix} A\sin\omega t + B\cos\omega t \\ \text{or} \\ Ce^{+i\omega t} + De^{-i\omega t} \end{matrix} \tag{3.8}$$

The solutions in equation (3.8) include pre-factors A, B, C, and D that for now must only meet two stipulations, they are: (1) time independent and (2) have dimension of SI length units: m. The eigenfunctions also contain imaginary factor $i = \sqrt{-1}$, and define a quantity known as the *angular speed* ω of the harmonic oscillator

(SI unit: rad · s^{-1}). Inserting either of the solutions from equation (3.8) into equation (3.7) shows the angular speed to be a function of the spring constant and the reduced mass:

$$\omega = \sqrt{\frac{k}{\mu}} \tag{3.9}$$

The solutions in equation (3.8) can also be written in forms such as: $x = A \sin 2\pi ft + B \cos 2\pi ft$ with the definition of frequency f (SI unit: s^{-1} or Hertz):

$$f = \frac{1}{2\pi} \sqrt{\frac{k}{\mu}} \tag{3.10}$$

The functions in equation (3.8) are the only valid solutions for eigenvalue equations with an operator involving second differentiation. The exponential and trigonometric forms can be inter-converted by expanding each in power series, which show that: $e^{\pm i\theta} = \cos \theta \pm i \sin \theta$. Only the exponential forms would be valid for operators involving first differentiation. We will later see these are required in certain quantum mechanical cases, particularly for eigenfunctions of both linear and angular momentum. But for now let us focus upon the real solution containing trigonometric functions. Its form is a natural description of vibrating matter in a wavelike fashion. It also seems that this solution is a combination of two independent sine and cosine waves. To borrow a term from chapter 1, the general solution for vibrating matter is a *superposition* of waves.

Further insight comes when the classical harmonic oscillator is treated as a *boundary value problem*, where the solutions in equation (3.8) are subject to constraints, known as *boundary conditions*. Let's suppose that our experiment is designed so that time measurement begins when the oscillator is in motion with the spring neither compressed nor stretched, so that the vibrating mass is at zero displacement. Mathematically this requires at $t = 0$ that $x = 0$. Because $\sin(0) = 0$ and $\cos(0) = 1$, the trigonometric function of equation (3.8) satisfies this boundary condition only if $B = 0$ (and, if we were instead using the imaginary solution, both $C = D = 0$). Under this constraint, the valid solution to the motion of the harmonically oscillating mass reduces to:

$$x = A \sin \omega t \tag{3.11}$$

The pre-factor A may now be given physical interpretation. It scales the sine function beyond its maximum of $+1$ and minimum of -1. In fact, A is the amplitude of the oscillating mass, or distance from the equilibrium position ($x = 0$) to the turning points at maximum spring stretch ($x = A$) and compression ($x = -A$). On a plot of the oscillator motion, A is the displacement to the positive and negative antinodes of the sine curve.

Philosophically speaking, the problem's general solution provided by equation (3.8) is omnibus, serving the sole purpose of including every viable mathematical solution to the problem. As long as we are not looking, equation (3.8) is throwing the kitchen sink at the problem in a *superposition of states* form. When boundary conditions are applied, we are taking a peek, with a prejudice that a particular outcome is expected. This causes the destruction of all other possibilities except the most overwhelmingly-likely result. Based on our stipulated requirements, the sine function survives as the only viable representation of the vibrating mass if it is to be located at $x = 0$ when we start our clock. The coefficient B has no possibility other than zero, to meet the constraint.

As a further example, consider the opposing captains of football teams meeting at the 50-yard line for the pre-game coin toss. The referee places a coin on the knuckle of his thumb, and flicks it into the air. As it flips over and over, it is in an indeterminate state—a superposition of both heads *and* tails. Even when trapped on his forearm with the downward-facing palm of his opposite hand, the coin remains in a superposition of states. It is not until he lifts his hand revealing the coin that a specific outcome is obtained. It would seem that events in the Universe are in some way dependent on the action of human observation. A familiar euphemism to this sentiment is the expression: 'if a tree falls in the forest, does it make a sound?'

Philosophical issues such as this fueled a spirited and sometimes contentious debate among some of the most influential physicists of the early 20th century. Much of the discussion focused on the deeper significance of *wavefunctions*, mathematical solutions containing information which determine probabilistic outcomes in the physical universe. These expressions arise as solutions to the differential equations describing the wave-like behavior of matter according to the rules of quantum mechanics. An 'un-observed' wavefunction possesses information for all states and possible outcomes, a condition known as *quantum superposition*. The action of human observation (via measuring properties or specifying boundary conditions) *collapses* the wavefunction to the state of most statistically-likely outcome.

This concept, suggested by Bohr, Born, and Heisenberg, is known as the *Copenhagen Interpretation*. Some scientists, often playing devil's advocate, proposed *gedanken* (*thought*) *experiments* to challenge this premise. The most-familiar of these is 'Schrödinger's Cat', where a feline is sealed in a box along with a radioactive isotope. If the isotope decays it emits lethal radiation, however the cat is in an indeterminate state of life or death unless the box is opened and the outcome is physically observed. A less whimsical challenge was posed in the EPR paradox (so-called after its authors Einstein, Podolsky, and Rosen) which suggested the Copenhagen Interpretation could not justify the limits of accuracy on measurable properties that two simultaneously-formed particles can possess unless they have the ability to communicate information faster than the speed of light. The particles are inextricably wed by what is known as *quantum entanglement*.

So far we have studied the harmonic oscillator for the purpose of introducing commonalities to the classical and quantum mechanical representation of matter, and to point out some of the conundrums that interpreting matter from a wave perspective introduces. We now look at the system energetics, which reveals an insurmountably fatal flaw in the classical representation of wave motion. Using equation (3.11), the harmonic oscillator kinetic energy is:

$$T = \frac{p^2}{2\mu} = \frac{1}{2}\mu \left| \frac{\mathrm{d}x}{\mathrm{d}t} \right|^2 = \frac{1}{2}\mu\omega^2 A^2 \cos^2 \omega t = \frac{1}{2}kA^2 \cos^2 \omega t \qquad (3.12)$$

Here the final identity follows from inserting the definition of angular speed from equation (3.9) into equation (3.12). Using equation (3.5) and equation (3.11) we can also find an expression for the potential energy:

$$V = \frac{1}{2}kx^2 = \frac{1}{2}kA^2 \sin^2 \omega t \qquad (3.13)$$

From simple trigonometric relations found in section B.2, the classical harmonic oscillator has total energy:

$$E_{\text{tot}} = T + V = \frac{1}{2}kA^2(\sin^2 \omega t + \cos^2 \omega t) = \frac{1}{2}kA^2 \qquad (3.14)$$

The right-hand side of equation (3.14) is time-independent, so the total energy is constant at all times. It is partitioned into kinetic and potential forms that have a relationship expressed in the *virial theorem*, which states a system subjected to potential of the form: $V(r) = c \cdot r^n$ has time-averaged kinetic and potential energies obeying:

$$2\langle T \rangle = n \cdot \langle V \rangle \qquad (3.15)$$

Using equation (3.15) and equation (3.5), the harmonic oscillator has average potential and kinetic energy related by: $\langle T \rangle = \langle V \rangle$. Figure 3.2 plots two cycles of the harmonic oscillator subject to boundary conditions $x = 0$ at $t = 0$. The process begins by stretching the mass to store an arbitrary amount of potential energy according to equation (3.5). While holding it stationary there is zero kinetic energy, so $E_{\text{tot}} = V$.

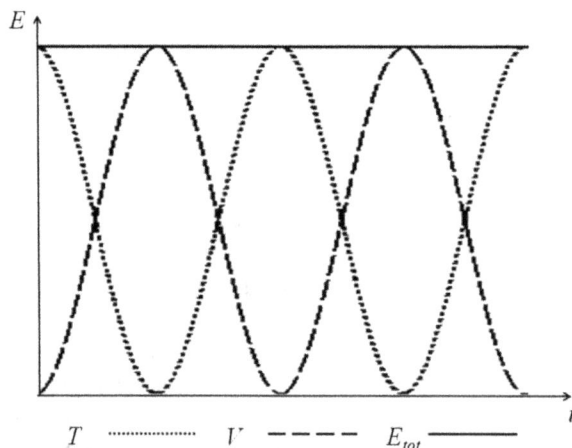

Figure 3.2. Plots of kinetic, potential, and total energy for the first two cycles of a classical harmonic oscillator.

The spring is released and the clock started when the mass is at its equilibrium position ($x = 0$). At this point (see equation (3.13)), the spring stores no potential energy so $E_{tot} = T$ and the system is at maximum velocity. As the mass drives past equilibrium, the spring compresses as the system slows and its kinetic energy is converted back to the stored form. This continues until all kinetic energy is converted to potential at the negative turning point and the cycle repeats.

The total energy being a function of the square of the oscillator amplitude does not clash with everyday experience. For instance, the energy (loudness) of sound is a function of the sound wave amplitude (the degree of compression and rarefication that the medium experiences). Another example is supplied by a pendulum, which uses gravity to supply the restoring force instead of a spring. We can envision a child on a swing, whose energy depends on the height she reaches above the ground at her turning points. An additional conclusion drawn from equation (3.14) is that the energy of the classical harmonic oscillator is *continuous*. Since the oscillator's amplitude A is allowed to take any real value desired (we can in theory pull the ideal spring out to any distance we wish), the energy E_{tot} can therefore have any real value.

3.2 Planck's approach to vibration

Now that we have a clear picture of the behavior of classical oscillation within the framework of Hamiltonian mechanics we turn to its shortcomings, particularly with regard to the ability of charges in vibrating matter to produce electromagnetic waves. If we enter this prejudiced by analysis of the last section, we would expect from equation (3.14) that matter would generate light with energy related to the magnitude of the oscillator force constant and the square of the oscillation amplitude. It would also seem this energy would have a continuous range. Both conjectures prove to be incorrect. The mathematics that resolved these issues posed an even more intriguing question. Were the shortcomings attributable to our understanding of waves or to our perception of matter?

At the turn of the 20th century, Rayleigh and Jeans used the results that we have so far proposed—a continuous energy harmonic oscillator model—to describe electromagnetic waves radiating from a black body. Their development relied upon classical thermodynamic arguments. According to the *principle of equipartition of energy*, temperature T produces an average energy of $\frac{1}{2}kT$ each for kinetic and potential contributions to the oscillator energy. The physical constant k used here is unfortunately not the spring force constant, but is now Boltzmann's constant, the ideal gas constant per particle (see the unit definitions of appendix A). The total energy per unit area is either referred to as the black body energy density or the *spectral brightness* (B). If the surface area is taken as the square of an oscillator's wavelength, then using $\lambda = c/\nu$:

$$B \propto kT\frac{\nu^2}{c^2} \tag{3.16}$$

From dimensional analysis, B has units of $N \cdot m^{-1}$ ($kg \cdot s^{-2}$) which is the dimensionality of an oscillator's force constant (unfortunately also represented by

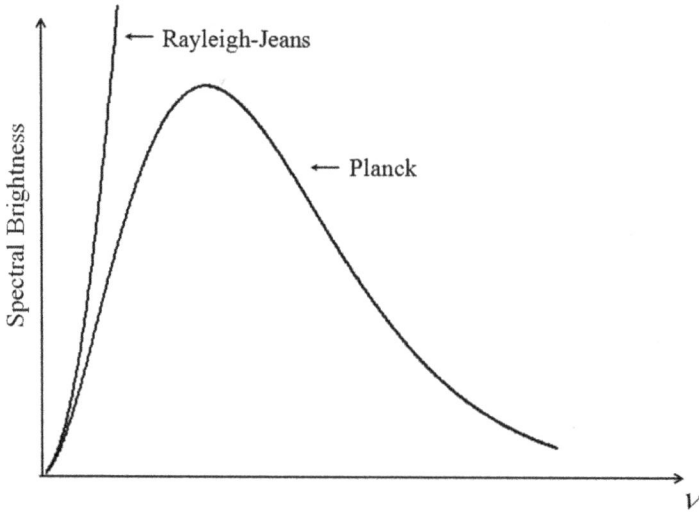

Figure 3.3. Comparing the Planck and Rayleigh–Jeans formulations of spectral brightness.

the symbol k). Multiplying B by the dimensionality of square amplitude does in fact give energy units as equation (3.14) stipulates. Equation (3.16) reproduces experimental results very accurately at low frequency (long wavelength), but as is depicted in figure 3.3 infinitely diverges beyond the visible range. Scientists therefore melodramatically referred to this as the *ultraviolet catastrophe*.

Around about the same time (actually, a little before) Max Planck studied the same problem but employed a *statistical* rather than classical thermodynamic treatment. He required oscillator energies distributed in *discrete*, evenly-spaced levels with values proportional to integer multiples of the *frequency* of oscillation:

$$E_n = nh\nu \quad n = 0, 1, 2, \ldots \tag{3.17}$$

The proportionality constant h in equation (3.17) has units of angular momentum, and is now known as Planck's constant. This value, which turns out to be inherent to all quantum mechanical phenomena, is often referred to as the *quantum of action*.

As we will ultimately learn in chapter 9, evenly-spaced vibrational energy levels are only strictly valid for parabolic potentials like that shown in figure 3.2. Real systems such as diatomic molecules have anharmonic potentials (see section 9.2) in which subsequent energy levels become increasingly closer in energy until reaching their dissociation limit. Typical spacing between vibrational energy levels is on the order of 10^{-20} J. Although this seems an inconsequential if not negligible amount of energy, it is actually intermediate in the quantum hierarchy. For instance, quantized rotational level spacings are on the order 10^{-21} J, while electronic levels: 10^{-19} J. Borrowing from statistical mechanics Planck's oscillators are summed into a vibrational partition function Z^V in the following fashion:

$$Z^V = \sum_{n=0}^{\infty} e^{-nh\nu/kT} \tag{3.18}$$

The nth energy level is called a *microstate* of the system, and its exponential term is a *Boltzmann factor*, representing the population of that state relative to the zeroth-level state. As equation (3.18) shows, Boltzmann factors are summed to give the partition function.

An analytical expression for the vibrational partition function results from applying the power series expansion: $1 + x + x^2\ldots = 1/(1 - x)$ with $x = e^{-h\nu/kT}$:

$$Z^V = \frac{1}{1 - e^{-h\nu/kT}} \tag{3.19}$$

In statistical mechanics, partition functions represent the number of microstates available to be populated subject to available ambient energy kT. Using a typical vibrational spacing of 10^{-20} J, we find from equation (3.19) that $Z^V(100\text{ K}) = 1.001$, $Z^V(300\text{ K}) = 1.10$, $Z^V(1000\text{ K}) = 1.94$, and $Z^V(2000\text{ K}) = 3.29$. The take-home is that at typical temperatures, only the ground vibrational state is significantly populated. If instead the partition function is numerically summed using equation (3.19) including only the first eight energy levels, we obtain: that $Z^V(100\text{ K}) = 1.001$, $Z^V(300\text{ K}) = 1.09$, $Z^V(1000\text{ K}) = 1.93$, and $Z^V(2000\text{ K}) = 3.11$. Hence the summations converge very rapidly when $h\nu \ll kT$. The fact that only the first one or two vibrational levels are significant contributors to Z^V is also justification for using a harmonic potential even for real systems at normal temperatures, as the rapid convergence of the sum occurs before anharmonicities become of major importance.

The energy contribution of the oscillators is found from the partition function using the statistical mechanical recipe:

$$E = kT^2 \frac{1}{Z^V} \frac{\partial Z^V}{\partial T} \tag{3.20}$$

It should be noted that equation (3.20) is in actuality an eigenvalue equation: $\hat{O}Z = cZ$ with operator: $\partial/\partial T$ and eigenfunction: Z^V. In a very short period of time we have seen that eigenvalue problems have applications in classical mechanics and statistical mechanics, and will soon enough be used in quantum mechanics as well. From this perspective we can also think of a partition function as a thermodynamic wavefunction. Using equation (3.19) and a little manipulation, it is straightforward to verify that equation (3.20) has solution:

$$E = h\nu \frac{e^{-h\nu/kT}}{1 - e^{-h\nu/kT}} \tag{3.21}$$

Upon multiplying the numerator and denominator above by $e^{+h\nu/kT}$. Planck's spectral brightness formula then results from dividing E by the oscillator surface area:

$$B \propto \frac{h\nu^3}{c^2} \frac{1}{e^{h\nu/kT} - 1} \tag{3.22}$$

As expected, this has the same dimensionality as equation (3.16).

When Planck's constant h was empirically-fit to experimental data, this expression was able to reproduce experimental spectral brightness of a black body radiator for all frequencies. In the visible range and below, both Planck and the Rayleigh–Jeans expressions mimic experiment. Although they appear to be significantly different, it is easy to show that in these frequencies where $h\nu \ll kT$ that expanding the exponential in a power series and neglecting terms quadratic and above, Planck's equation simplifies to the Rayleigh–Jeans form in equation (3.16).

Modeling oscillating matter to produce electromagnetic radiation by employing energy proportional to oscillator frequency instead of amplitude conflicted with classical interpretation. Planck's bold assumption of discrete energy oscillators would quickly be used by the physics community to resolve other troubling experimental–theoretical discrepancies. His approach was seized upon by Einstein to first interpret the heat capacity of low temperature solids. He then applied the same approach to explain the *photoelectric effect*, the phenomenon of shining light on a thin metal foil to cause the ejection of electrons. If electromagnetic radiation interacts with electrons classically, light with energy proportional to wave amplitude (as in equation (3.14)) would eject electrons after reaching a particular level of brightness. It was experimentally determined that, independent of intensity, electrons did not begin to be ejected until the light reached a threshold *frequency*, which Einstein called the *work function* of the material. Increasing light frequency above the threshold caused a linearly proportional increase in the kinetic energy of ejected electrons. Much like he did in resolving the heat capacity issue, Einstein's theoretical model of the photoelectric effect made use of Planck's expression for light energy of equation (3.17).

A couple of decades later, Wolfers and Lewis coined the term *photon* for the frequency-dependent packet of energy carried by light. According to equation (3.17), a green light photon of wavelength 540 nm (frequency 560 THz) possesses energy 3.7×10^{-19} J. It is no wonder this miniscule energy would be hard to detect, and only tangibly influence matter of incredibly small mass or dimension. Light intensity does play a role, but instead of proportionality to photon energy, it is to the *number* of photons a light beam carries, hence the *number* electrons ejected, not their energy. For instance a mole of green light photons have energy 220 kJ, which is the same amount released upon complete combustion of 4 g ($\approx$6 L) of methane gas.

One more point regarding the form of equation (3.17). It can equivalently be represented using the speed of light and its wavelength as: $E = hc/\lambda$, but both c and λ vary depending on the type of matter comprising the medium in which light propagates. As electromagnetic radiation enters matter with index of refraction n, its speed slows and its wavelength shortens according to the expressions $c' = c/n$ and $\lambda' = \lambda/n$. Even though these effects cancel and the photon energy is independent of the medium, it is more satisfactory to express energy in terms of photon frequency, which is invariant to the matter comprising a medium.

In summary, it now seems apparent that light (electromagnetic radiation in general) packs energy proportional to its frequency not its amplitude. Furthermore, this energy does not interact with matter continuously but in discrete bundles referred to as *quanta*. In a variety of examples, with even more to come, modeling the interaction of matter and electromagnetic waves inevitably involves the quantum of action h. These conclusions initially required our understanding of light to be altered as it encountered matter. However, after a few decades quantum mechanics would flip the perspective. As a consequence, matter exhibited properties previously reserved for light waves. Ultimately, what would evolve is the concept of *wave–particle duality*, in which the role of matter and waves depend on the situation and method of observation.

IOP Concise Physics

What's the Matter with Waves?
An introduction to techniques and applications of quantum mechanics
William Parkinson

Chapter 4

Rotating matter

4.1 Analysis of classical rotational motion

Our focus switches to the internal motion caused by simultaneous changes in the angular (not radial) displacement of matter, subject to maintaining its center of mass. It is easy to envision that rotational motion at a constant speed is inherently periodic. You need look no further than the phases of the Moon or the ocean's tides, the four seasons or circadian rhythms of life to appreciate this fact. As before, we will begin the study of rotation by placing ourselves in the mindset of the physics community at the turn of the 20th century. For simplicity we consider rotation with no angular acceleration. The rotor is orbiting a central point at a fixed rate during each of its cycles. It is in fact only then that the rotor is harmonically oscillating.

We begin by designing a system according to the discussion in chapter 2, with mass μ rotating counter-clockwise in the 2-dimensional Cartesian x–y plane about the fixed central point of the coordinate system origin (refer to figure 4.1). First note that μ will remain in circular motion only if some contact force or field potential is acting on the mass, otherwise it will fly off with linear kinematics. For example throwing fast-pitch softball requires the pitcher to apply tension on the ball via her arm during the wind-up. Similarly, a road exerts frictional force on the tires to cause an automobile to execute a curve. The same effect can occur for non-contact forces, such as satellites maintaining circular motion due to gravity, and the movement of charged particles as a result of the influence of either electric or magnetic fields. The mathematical development for atoms in chapter 11 will detail the nature of rotation as related to a force which compels this motion.

The rotor has angular momentum: $\vec{L} = \vec{r} \times \vec{p}$ with a z component of magnitude:

$$L_z = xp_y - yp_x = \mu\left(x\frac{\mathrm{d}y}{\mathrm{d}t} - y\frac{\mathrm{d}x}{\mathrm{d}t}\right) \tag{4.1}$$

doi:10.1088/978-1-6817-4577-0ch4

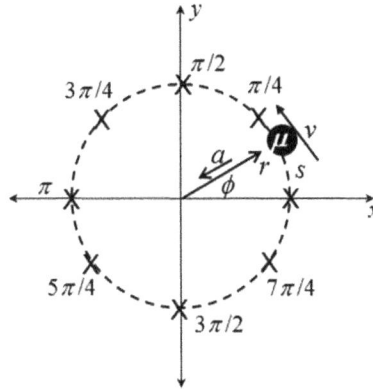

Figure 4.1. The rotational coordinate frame.

(this is in fact the only non-zero component of angular momentum in the system). To solve the differential equation, it is much more convenient to employ a spherical polar coordinate system as defined in figure 4.2, allowing Cartesian coordinates to be transformed using the relationships: $x = r \sin \theta \cos \phi$ and $y = r \sin \theta \sin \phi$. The angular momentum in this representation is:

$$
\begin{aligned}
L_z = {} & \mu r \sin \theta \cos \phi \left(r \cos \theta \sin \phi \frac{d\theta}{dt} + r \sin \theta \cos \phi \frac{d\phi}{dt} + \sin \theta \sin \phi \frac{dr}{dt} \right) \\
& - \mu r \sin \theta \sin \phi \left(r \cos \theta \cos \phi \frac{d\theta}{dt} - r \sin \theta \sin \phi \frac{d\phi}{dt} + \sin \theta \cos \phi \frac{dr}{dt} \right)
\end{aligned}
\tag{4.2}
$$

At first glance this does not seem convenient, but simplifications from this choice result from two conditions inherent to the system design. First the rotor is confined to the x–y plane, so the polar angle is a constant value of $\theta = 90°$ for all time. Secondly, as discussed in chapter 2, we use a rigid rotor so that r is also fixed. Imposing these conditions, combining terms, and a little trigonometry reduces equation (4.2) to the following function of the radial distance r and the rate of change of the azimuthal angle ϕ:

$$
L_z = \mu r^2 \frac{d\phi}{dt}
\tag{4.3}
$$

We identify the moment of inertia I of the rigid rotor:

$$
I = \mu r^2
\tag{4.4}
$$

and also note the rate of change of the azimuthal angle ϕ is the angular speed of the rotor. This name was previously associated with periodic vibrational motion in chapter 3 and possesses the same units of rad · s^{-1}, but in this instance is defined:

$$
\omega = \frac{d\phi}{dt}
\tag{4.5}
$$

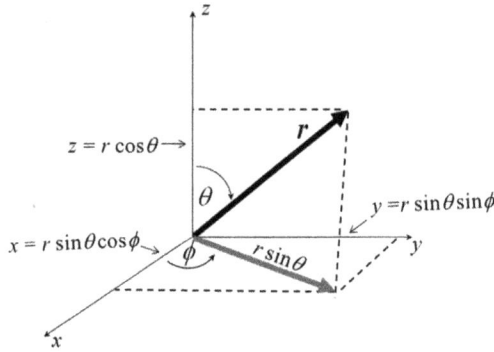

Figure 4.2. The spherical polar coordinate system.

Using quantities defined from equations (4.4) and (4.5), the rotor angular momentum is:

$$L_Z = I\omega \tag{4.6}$$

Equation (4.6) is formulated in much the same way as linear momentum $p = mv$ with I playing the role of mass and ω velocity. Just like its linear counterpart, angular momentum is a conserved quantity. If r (hence I) is allowed to change, ω compensates in the opposite sense to maintain the initial L value. For example, a skater spins more rapidly when she tucks in her arms. We simplify our rotor problem by imposing the rigid rotor condition for this reason.

If we further introduce angular acceleration α as the rate of change of angular speed, we have a complete set of functions describing rotational kinematics that bear a striking resemblance to their counterparts in the linear frame, including rotational force, or torque: $\tau = I\alpha$. Just as linear force is the rate of change of linear momentum, for a rigid rotor (with implicit constant r) torque can be expressed as:

$$\tau = \frac{\mathrm{d}L}{\mathrm{d}t} \tag{4.7}$$

Before we moving on it is important to note since ω is free to take any value, the rotor has continuous angular momentum.

A similar approach can be taken for the rotor's kinetic energy in a 2-dimensional Cartesian coordinate system:

$$T = \frac{1}{2}\mu\left(\frac{\mathrm{d}x}{\mathrm{d}t}\right)^2 + \frac{1}{2}\mu\left(\frac{\mathrm{d}y}{\mathrm{d}t}\right)^2 \tag{4.8}$$

Again transforming to spherical polar coordinates gives:

$$T = \frac{1}{2}\mu\left(-r\sin\theta\sin\phi\frac{d\phi}{dt} + r\cos\phi\cos\theta\frac{d\theta}{dt} + \sin\theta\cos\phi\frac{dr}{dt}\right)^2$$
$$+ \frac{1}{2}\mu\left(r\sin\theta\cos\phi\frac{d\phi}{dt} + r\sin\phi\cos\theta\frac{d\theta}{dt} + \sin\theta\sin\phi\frac{dr}{dt}\right)^2$$

(4.9)

Simplifying for a time-invariant r at a fixed polar angle of 90° gives:

$$T = \frac{1}{2}\mu r^2\left(\frac{d\phi}{dt}\right)^2 = \frac{1}{2}I\omega^2$$

(4.10)

It is noted that equation (4.10) is consistent in form with linear kinetic energy $T = 1/2mv^2$ where I plays the role of mass and ω velocity. In addition, we see by comparing equations (4.10) and (4.6) that rotational kinetic energy can be expressed in terms of angular momentum as: $T = L^2/2I$. Equation (4.10) is also consistent with the description of rotational motion in chapter 2 as a change in *direction* (via the azimuthal angle) of the internal motion vector, not its magnitude.

Recalling results from chapter 3, a classical harmonic oscillator has kinetic, potential, and total energies found to be directly proportional to the square of amplitude of oscillation A (see equations (3.12), (3.13), and (3.14)) . From equation (4.10) we see that T is directly proportional to I which, according to equation (4.4), is directly proportional to r^2. This radial distance plays the same role in a rotor as amplitude of displacement in a vibrator, and as a consequence both show quadratic relation to energy. Finally, it is important to note that equation (4.10) shows rotational kinetic energy to be a continuous function, as μ, r^2, or ω can each in theory take any real value.

Referring again to figure 4.1, suppose a clock is started as μ rises above the abscissa. It moves in a circular path of arc length s while sweeping out angle ϕ of magnitude:

$$\phi = \frac{s}{r}$$

(4.11)

If the mass continues until arc length s equals the circle's circumference, angle ϕ reaches 360°. Alternatively, the angle can be measured by inserting the definition of circumference in equation (4.11):

$$\phi = \frac{2\pi r}{r} = 2\pi$$

(4.12)

Here we have introduced the connection between angular displacement in degrees and its SI counterpart, the *radian*. As μ executes counter-clockwise circular motion, figure 4.1 shows the angular displacement in radians for selected points of the first cycle. The value of ϕ at those points on each subsequent rotational cycle is found by adding an additional 2π per cycle.

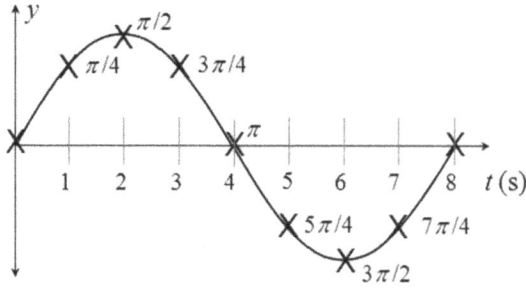

Figure 4.3. Plot of the rotational motion of the rigid rotor.

Suppose that the angular speed is set at a constant value of $\omega = \pi /4$ rad $\cdot$ s^{-1}. Plotting the mass displacement above and below the abscissa against time generates the curve shown in figure 4.3. The y-axis displacements at the ϕ values denoted in figure 4.1 are included in figure 4.3. As we learned with vibrational motion in chapter 3, matter undergoing rotation has mathematical description that is also wave-like. Furthermore, if the angular speed is constant, the motion of μ is periodic. If the time axis were to be extended, the pattern for subsequent 2π, 4π ... cycles of y displacement would repeat every 8.0 seconds. This time interval is the *period* of rotation (T), defined as the time required to complete each cycle of motion. The period is calculated as the angular displacement of a complete cycle divided by the constant angular speed:

$$T = \frac{2\pi}{\omega} = \frac{2\pi \text{ rad}}{\pi/4 \text{ rad} \cdot \text{s}^{-1}} = 8.0 \text{ s} \tag{4.13}$$

(It is an unfortunate consequence of convention that the symbol T is used to represent temperature, a period of harmonic motion, and kinetic energy of a mass.) Analogous to vibrations in chapter 3, the inverse of T represents the number of rotational cycles completed per second, or frequency f:

$$f = \frac{1}{T} = \frac{\omega}{2\pi} = 0.125 \text{ s}^{-1} = 0.125 \text{ Hz} \tag{4.14}$$

Our previous discussion pointed out the striking similarities between equations for rotational and linear kinematics, momentum, force and energy. We now show the connection between angular and linear speed. We begin with a radial vector extending from the coordinate origin to μ as in figure 4.1. This vector is expressed in spherical polar coordinates as:

$$\vec{r} = r \cos \phi \vec{i} + r \sin \phi \vec{j} \tag{4.15}$$

In equation (4.15) the x and y components of the unit vector assume polar angle θ maintains a time-independent value of 90°. With help from equation (4.5), the linear velocity is:

$$\vec{v} = \frac{d\vec{r}}{dt} = -r\omega \sin \phi \vec{i} + r\omega \cos \phi \vec{j} \tag{4.16}$$

$$\vec{r} = r\cos\phi\,\vec{i} + r\sin\phi\,\vec{j} \qquad \vec{v} = -r\omega\sin\phi\,\vec{i} + r\omega\cos\phi\,\vec{j}$$

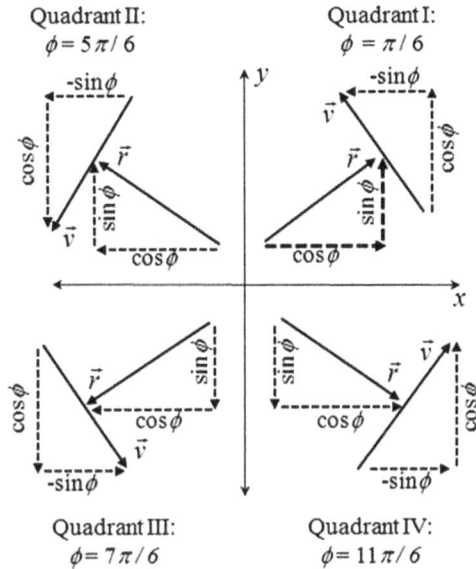

Figure 4.4. Examples of relative orientation for the linear velocity and radial vectors at angles in each Cartesian quadrant (using: $r = 1$ m and $\omega = 1$ rad · s^{-1}).

Figure 4.4 gives a detailed insight into the relationship between the radial vector $\vec{r}$ and the linear velocity vector $\vec{v}$. Examining the behavior of sine and cosine in each quadrant, we see that the velocity vector is tangent to the circular trajectory and perpendicular to the radial vector at all points of rotation. Linear speed is found from the velocity vector magnitude to be:

$$v = \sqrt{(-r\omega\sin\phi)^2 + (r\omega\cos\phi)^2} = r\omega \qquad (4.17)$$

which shows the tangential linear speed: v is directly proportional to both the angular speed: ω and the rotational *moment arm*: r. For instance, all riders enjoying a merry-go-round are moving at the same angular speed, as no-one is catching or passing anyone else no matter how far out towards the edge they sit. However an individual a distance r from the center of rotation would fly off the ride with ½ the linear speed of a person out a distance $2r$ (because of the shorter moment arm, this individual also possesses ¼ the rotational and linear kinetic energy).

We are now in position to examine the linear acceleration mass μ experiences, known more formally as the *centripetal acceleration*. This is found from the rate of change of equation (4.16). Using equation (4.5), along with the facts that the rotor is moving: (1) at constant angular speed: ω and (2) with fixed radial distance: r leads to:

$$\vec{a} = \frac{d\vec{v}}{dt} = \frac{d^2\vec{r}}{dt^2} = -r\omega^2\cos\phi\vec{i} - r\omega^2\sin\phi\vec{j} = -\omega^2\vec{r} \qquad (4.18)$$

Equation (4.18) illustrates the striking similarities between harmonic vibrational and rotational motion. In fact the acceleration of the harmonic vibrator has this exact same form, as can be seen from the second time derivative of equation (3.11). Centripetal acceleration of a rigid rotor is directed in the direction opposite the radial vector (see figure 4.1) and has magnitude:

$$a = \sqrt{(-r\omega^2 \cos \phi)^2 + (-r\omega^2 \sin \phi)^2} = r\omega^2 \qquad (4.19)$$

As Newton's Second Law requires, centripetal acceleration is induced by a linear *centripetal force* that constantly drags μ inwards towards it center of rotation. As is true of all linear forces, centripetal forces cause acceleration along their direction of action. Centripetal force is expressed in terms of linear speed by combining Newton's Second Law with the results of equations (4.17) and 4.19):

$$F = \mu a = -\mu \frac{v^2}{r} \qquad (4.20)$$

4.2 Bohr's approach to rotation

About a decade after Planck, and subsequently Einstein, used vibrating matter with a discrete, frequency-dependent energy distribution as a model of interaction involving electromagnetic waves, Bohr adapted the same approach for the rotational motion of an electron around a nucleus. This technique would not only prove successful in explaining how the hydrogen atom interacted with electromagnetic waves, but moreover was able to predict basic experimental properties of hydrogen in the absence of light. With Bohr's success, it became more apparent that inconsistencies between theory and experiment with regard to interpreting light–matter interaction were equally the fault of our perception of matter as it was with waves.

Applying classical rotational mechanics to the behavior of electrons in atoms was a particular source of consternation for early 20th century physicists. In 1909, Rutherford, Geiger and Marsden conducted their famous 'gold foil' experiment, designed to prove or disprove the 'plum pudding' model of the atom suggested by rival Thomson (who is routinely portrayed in essentially all freshman chemistry texts as 'the guy who got it wrong'. This is a complete disservice to Thomson's experimental acumen. In 1897 his lab had discovered the electron, which proved the existence of subatomic particles and motivated attempts to explain how neutral matter could intrinsically contain both positive and negative charges). Rutherford subsequently proposed the 'planetary model' in which atoms are envisioned as massive, positively charged 'nuclei' (a term coined by Faraday in 1844, by the way) orbited by significantly smaller negatively charged electrons confined to their motions by Coulombic centripetal forces.

Although this was a reasonable explanation of experimental observation, the planetary model suffered two fatal flaws. Even if moving at constant angular speed, orbiting electrons are accelerated inward by their attraction to the nucleus. It was understood from electrodynamics that accelerating charges must emit electromagnetic radiation, so all atoms—whether stimulated in some fashion or not—should possess a characteristic, detectable light frequency. No such radiation could be detected from un-perturbed atoms. Secondly, a radiating electron expends energy, which should cause it to eventually spiral inward and crash into the nucleus.

In 1913, Bohr addressed these shortcomings by modeling a negative charge orbiting a center of positive charge in the following fashion. He began with the expression for Coulombic electrostatic force in the form:

$$F = -\frac{Ze^2}{(4\pi\varepsilon_0)r^2} \tag{4.21}$$

Equation (4.21) represents the charge on an electron by $-e$ and the charge of an arbitrary nucleus as $+Ze$, where Z is the number of protons, or atomic number of the nucleus. It also contains the vacuum permittivity of space $4\pi\varepsilon_0$ (see appendix A), a quantity required for dimensional analysis in SI units. He then used equation (4.20) to equate the centripetal force an electron experiences:

$$-\mu\frac{v^2}{r} = -\frac{Ze^2}{(4\pi\varepsilon_0)r^2} \tag{4.22}$$

Combining equations (4.4), (4.6) and (4.17) he expressed the angular momentum of the electron in terms of its linear velocity:

$$L = \mu r v \tag{4.23}$$

and inserted this into equation (4.22):

$$-\frac{L^2}{\mu r^3} = -\frac{Ze^2}{(4\pi\varepsilon_0)r^2} \tag{4.24}$$

Instead of assuming L was a classically-continuous function, Bohr chose the approach of Planck and Einstein. He restricted angular momentum of the electron to only integer amounts of the quantum of action h per each complete rotational cycle of 2π radians:

$$L = \frac{Nh}{2\pi} \quad N = 1, 2, 3, \ldots \tag{4.25}$$

In periodic systems, the quantum of action: h is often normalized per complete angular cycle. This form of Planck's constant takes the form: $\hbar = h/2\pi$. Bohr's expression thus becomes:

$$\frac{N^2\hbar^2}{\mu r^3} = \frac{Ze^2}{(4\pi\varepsilon_0)r^2} \tag{4.26}$$

This expression can be solved for the distance of an electron from the nucleus as a function of energy level: N, atomic number: Z, and all other physical constants collected together. Using these values from appendix A we obtain:

$$r = \frac{N^2\hbar^2(4\pi\varepsilon_0)}{Ze^2\mu} = \frac{N^2}{Z}a_0 = \frac{N^2}{Z}5.29 \times 10^{-11}\,\text{m} = \frac{N^2}{Z}0.529\,\text{Å} \qquad (4.27)$$

When determined for the lowest energy level ($N = 1$) for a single proton nucleus ($Z = 1$), the value of r is the correct average distance between an electron and proton in the hydrogen atom ground state. The collections of physical constants defined as a_0 is now known as the *Bohr radius*. (Please see appendix A if you are interested in a discussion of the units used in atomic physics.)

The expression $E = h\nu$ was not used in the derivation of equation (4.27). In other words the Bohr radius is not a result of light acting as a particle, packing energy in discrete photon bundles. It arises from treating electrons as possessing periodically recurring, or wave-like, properties. Secondly, no additional parametric information is needed to find the distance between an electron and proton than a collection of fundamental constants of the Universe. Predicting the radius of a hydrogen atom was not the only success of Bohr's model. To calculate total energy, we first find an expression for kinetic energy as a function of angular momentum in the following fashion:

$$T = \frac{L^2}{2I} = \frac{N^2\hbar^2}{2\mu r^2} = \frac{N^2\hbar^2 e^4 \mu^2 Z^2}{2\mu N^4 \hbar^4 (4\pi\varepsilon_0)^2} \qquad (4.28)$$

The final identity arises from inserting r from equation (4.27). As for the potential energy V, a formula is obtained from Hamilton's equation from chapter 2 (equation (2.2)) using Coulomb's Law for electrostatic force between charged particles (equation (4.21)). This same technique was done to find the harmonic oscillator potential in equation (3.5). For a hydrogen atom the potential is then:

$$V = -\int F \cdot dr = -\frac{Ze^2}{(4\pi\varepsilon_0)r} = -\frac{Z^2 e^4 \mu}{(4\pi\varepsilon_0)^2 \hbar^2 N^2} \qquad (4.29)$$

where we have again used the definition of r from physical constants in equation (4.27). A common denominator is found for equations (4.28) and (4.29), which are then added to give the total energy:

$$E_{\text{tot}} = T + V = \frac{\mu e^4 Z^2}{2(4\pi\varepsilon_0)^2 \hbar^2 N^2} - \frac{2\mu e^4 Z^2}{2(4\pi\varepsilon_0)^2 \hbar^2 N^2} \qquad (4.30)$$

From equation (4.29) we note the potential energy is of the form: $V(r) = c \cdot r^{-1}$. Then in accordance with the Virial Theorem (See chapter 3, equation (3.15)), the above expression shows that $2\langle T \rangle = -\langle V \rangle$. Simplifying equation (4.30) leads to:

$$E_{\text{tot}} = \frac{\mu e^4 Z^2 - 2\mu e^4 Z^2}{2(4\pi\varepsilon_0)^2 \hbar^2 N^2} = -\frac{1}{2}\frac{Z^2}{N^2}\frac{\mu e^4}{\hbar^2(4\pi\varepsilon_0)^2} = -\frac{1}{2}\frac{Z^2}{N^2}\frac{\hbar^2}{\mu a_0^2} = -\frac{1}{2}\frac{Z^2}{N^2}E_h \qquad (4.31)$$

The energy unit defined in equation (4.31) is known as the *Hartree*, which is evaluated using values found in appendix A: $E_h = 4.36 \times 10^{-18}$ J $= 27.2$ eV $= 628$ kcal mol^{-1}.

Bohr's approach was the first to demonstrate that atoms theoretically possessed a discrete energy distribution. The presence of N quantizes energy levels, successively increasing their energy (E_{tot} becomes less negative with increasing N). The existence of energy levels was verified through the observation of light emitted by a stimulated atom. These spectra could be resolved into emission 'lines' occurring at very specific wavelengths, as opposed to continuous emission bands. Arguably the most significant achievement of Bohr's atom was its ability to correctly reproduce the emission spectra of hydrogen gas not only in the visible, but also in various regions of the ultraviolet and infra-red as well.

Emission spectra can be calculated from the Bohr model and equation (4.31) by finding the difference between energy levels:

$$\Delta E = -\frac{1}{2}\frac{Z^2\mu e^4}{\hbar^2(4\pi\varepsilon_0)^2}\left(\frac{1}{N_2^2} - \frac{1}{N_1^2}\right) \tag{4.32}$$

Electronic transitions are induced either by absorption ($N_2 > N_1$) or emission ($N_2 < N_1$) of electromagnetic energy according to what is known as the *Bohr frequency condition*:

$$\Delta E = h\nu = \frac{hc}{\lambda} \tag{4.33}$$

Spectroscopists often rewrite the above as: $\Delta E = hc\tilde{\nu}$ using wavenumbers: $\tilde{\nu} = 1/\lambda$ expressed in units of inverse centimeters (cm^{-1}). This quantity represents the number of complete waves which fit in a length of 1 cm. Using the Bohr frequency condition in equation (4.32) gives the following expression for the transition between energy levels:

$$\tilde{\nu} = Z^2\frac{2\pi^2\mu e^4}{h^3c(4\pi\varepsilon_0)^2}\left(\frac{1}{N_1^2} - \frac{1}{N_2^2}\right) = Z^2R\left(\frac{1}{N_1^2} - \frac{1}{N_2^2}\right) \tag{4.34}$$

Note in order that $\tilde{\nu}$ have units of cm^{-1}, c must be expressed in cm $\cdot$ s^{-1}. Equation (4.34) introduces the quantity R, known as the Rydberg Constant, which when calculated from values in appendix A is: $R = 109\ 700$ cm^{-1}.

Spectra for the hydrogen atom using equation (4.34) are remarkably accurate. The model can be used to predict the energy required to remove an electron from its ground state ($N_1 = 1$) to the ionization limit ($N_2 = \infty$), from which we obtain: $E = hc\tilde{\nu} = 13.6$ eV. This is the well-known experimental ionization energy of atomic hydrogen. In 1885, Balmer noted the wavelength of the four lines of hydrogen's visible emission spectrum could be empirically fit to the following expression:

$$\lambda = B\left(\frac{n^2}{n^2 - 2^2}\right) \qquad n = 3,\ 4,\ 5,\ \text{or } 6 \tag{4.35}$$

Table 4.1. Comparing hydrogen atom experimental emission spectral wavelengths to those calculated from the Bohr model.

N_1	N_2	λ (theoretical)	λ (experimental)[1]
Lyman Series (ultraviolet transitions)			
1	2	121.57 nm	121.6 nm
1	3	102.57 nm	102.6 nm
1	4	97.254 nm	97.0 nm
1	∞	91.176 nm	91.2 nm
Balmer Series (visible transitions)			
2	3	656.47 nm	656.3 nm
2	4	486.27 nm	486.1 nm
2	5	434.17 nm	434.1 nm
2	∞	364.70 nm	365.0 nm
Paschen Series (infra-red transitions)			
3	4	1875.6 nm	1875.1 nm
3	5	1282.2 nm	1281.8 nm
3	6	1094.1 nm	1093.8 nm
3	∞	820.58 nm	822.0 nm

[1] Taken from: Herzberg, *Spectra of Diatomic Molecules*, (1950) Van Nostrand Reinhold Co., New York, NY.

provided that the value $B = 364.5$ nm was employed. Balmer's formula was only capable of reproducing hydrogen's emission spectrum in the visible range. In 1888 Rydberg was able to formulate a generalized expression for hydrogen in the form shown on the right-hand side of equation (4.34), with the value $Z = 1$, and the empirical constant $R = 109,700$ cm^{-1}. He arrived at this number by fitting data, not from first principle application of physical constants as was done by Bohr. Either Rydberg's or Bohr's formula give the same results as equation (4.35) of Balmer when N_1 is set equal to 2. Equation (4.34) was also able to successfully predict hydrogen's ultraviolet ($N_1 = 1$, $N_2 = 2, 3, 4...$) and infra-red ($N_1 = 3$, $N_2 = 4, 5, 6...$) emission spectra as well. Comparison of experimental emission spectra and values predicted by Bohr's model are provided in table 4.1.

Bohr's model of the atom was a pivotal success to proponents of the new quantum theory. However it lacks general applicability beyond one electron atoms. It can be successfully applied to Li^{+2} or Au^{+78}, but not to neutral He, let alone H_2^{+1}. It would be another decade before a more robust theory would be developed with the capability to tackle such systems. These refinements are introduced in upcoming chapters.

IOP Concise Physics

What's the Matter with Waves?
An introduction to techniques and applications of quantum mechanics
William Parkinson

Chapter 5

Translating matter

5.1 Analysis of classical translational motion

Logic may suggest this part of the story be told prior to treating vibration or rotation but, in keeping with a theme of following the historic chronology of events, attention now turns to translation of matter. It is not a radical departure to picture wave character as matter rotates or vibrates. The periodic character of such motions readily lend themselves to a wave treatment. From the point of view of everyday experience, however, it seems unlikely that matter flying about would even be suspected of harboring any semblance of wavelike behavior. In that regard it is not surprising this is the last of matter's properties where wave behavior was proposed. It is therefore useful to study a case of matter in translation where we can take advantage of knowledge gained from studying periodic motion.

As a model system, consider a string stretched to an extent that it experiences some tension. The string has no particular defined length, but is attached at one end to a mechanical device which displaces it above and below its equilibrium position with a regular, repeating period T. This produces *transverse wave* motion in the string, meaning its matter undergoes a y-direction disturbance that propagates in the x-direction. Displacement of string particles in the vertical sense is described by the dual function:

$$y(x,\ t) = A \sin \frac{2\pi}{\lambda}(x - vt) \tag{5.1}$$

The mechanical oscillator is harmonic, producing a wave propagating at constant velocity: v (SI unit: $\mathrm{m \cdot s^{-1}}$), with displacement proportional to amplitude: A (SI unit: m), and wavelength: λ (SI unit: m). When exploring vibration in chapter 3 and rotation in chapter 4, the concept of wavelength did not play a significant role. Here it carries more relevance. The factor $2\pi/\lambda$ scales the function, ensuring that each wave has the same y-displacement every time an entire cycle of motion is completed.

Unlike harmonic oscillation of a vibrating mass (according to equation (3.11)), matter displacement is now a function of two variables. To understand the nature of this sine wave, consider tracking a particular positive antinode making its way down the string. As time progresses, x increases proportionally, by an amount depending on the constant: v, to remain at the antinode with value: $y(x,t) = A$. We must travel down the string at the same velocity as the pulse to maintain our coordinate frame. The expression: $x - vt$ is constant, so when differentiated:

$$\frac{dx}{dt} - v = 0 \qquad (5.2)$$

So that: $v = dx/dt$ is the *phase velocity* of the disturbance propagating through the matter and down the string.

Because x must increase with time, it can also be concluded that equation (5.1) describes a pulse travelling in the positive x-direction. If the wave were travelling in the negative x sense, equation (5.1) would use the function: $x + vt$. Fixing the value of t is the same as taking a snapshot of the string at a particular instant. The disturbance is no longer travelling, the product: vt is simply a constant phase factor shifting the sine wave origin, and we are just viewing matter displacement defining a fixed waveform: $y(x)$. Choosing a particular constant x value is equivalent to focusing attention on a particular section of matter in the string and watching it harmonically oscillate from positive to negative antinode: $y(t)$ according to equation (3.11). The fixed x-value is now acting as the phase constant, shifting the time at which the oscillator passes through equilibrium during its period of motion. To satisfy sticklers for mathematical rigor, there is one last detail about the form chosen for equation (5.1). This function is valid for the displacement boundary condition: $y(0,0) = 0$, or that the string is at its equilibrium position at the point where the wave is generated when the clock is started. To include the possibility of any starting point, the sine function should be the more general form: $x - vt + \phi$. Finally note that equation (5.1) can alternatively be expressed:

$$y(x, t) = A \sin(kx - \omega t) \qquad (5.3)$$

by using the definition: $k = 2\pi/\lambda$ (SI unit: m^{-1}) along with the relationship between wavelength, frequency, and wave velocity: $v = \lambda \times f$, and the angular speed (see chapter 4, equation (4.14)): $\omega = 2\pi f$.

Analysis of the phase velocity is aided by figure 5.1, showing a perturbed arc segment of length: l assumed to be travelling at a rate: v left to right down the string. The segment has *linear density*: μ_l (SI units: $kg \cdot m^{-1}$), a measure of its mass to length ratio. Each end of segment l is subjected to tangential tension F in opposed directions. Tension is typically denoted T, but since that variable has already been used three times: for symbolic representation of the period of harmonic motion, for temperature, as well as for kinetic energy in the Hamiltonian, F will be used to avoid any further confusion. In figure 5.1 the string disturbance is superimposed with a circle to highlight its relation to rotational kinematics.

The string segment has equal and opposite horizontal tension components ($\sum F_x = 0$), and reinforcing vertical tension components with total magnitude:

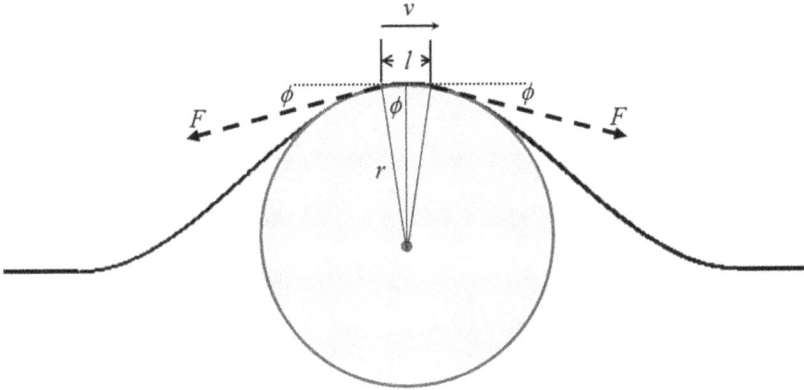

Figure 5.1. Force diagram for a disturbance of length Δl propagating through a string.

$(\sum F_y = 2F \sin \phi)$. The angle ϕ is assumed small enough that the approximation: $\sin \phi \cong \phi$ is valid. Using this simplification and borrowing from chapter 4, equation (4.11) with arc length: $s = l/2$ (refer to figure 5.1) gives a total force on the segment of:

$$F_{\text{tot}} = 2F\theta = 2F\frac{l/2}{r} = F\frac{l}{r} \tag{5.4}$$

Equation (5.4) represents the centripetal force experienced by the arc segment. This segment has mass: $m = \mu_l \cdot l$, so we can again borrow from knowledge of rotational motion gained in chapter 4, equation (4.20) to write:

$$F\frac{l}{r} = \frac{\mu_l \cdot lv^2}{r} \tag{5.5}$$

Equation (5.5) is solved for the phase velocity of the perturbation as it moves through matter down the string:

$$v = \sqrt{\frac{F}{\mu_l}} \tag{5.6}$$

The infinitesimal of kinetic energy in the horizontal direction can be determined using equation (5.6) along with the definition of linear density:

$$dT_x = \frac{1}{2}\mu_l \cdot dl_x v^2 = \frac{1}{2}F \cdot dl_x \tag{5.7}$$

To examine how energy is distributed in a segment of matter during vertical displacement, equation (5.3) is used. The kinetic energy is:

$$dT_y = \frac{1}{2}\mu_l \cdot dl_x \left|\frac{\partial y(x,t)}{\partial t}\right|^2 = \frac{1}{2}\mu_l \cdot dl_x \omega^2 A^2 \cos^2(kx - \omega t) \tag{5.8}$$

Extracting the vertical potential energy requires a little more imagination. First note a string segment undergoes extension in both dimensions:

$$dl = \sqrt{dx^2 + dy^2} = dx\sqrt{1 + \left(\frac{dy}{dx}\right)^2} \qquad (5.9)$$

For small values of x, the identity: $1 + x^2 \cong (1 + \frac{1}{2}x^2)^2$ can be used on equation (5.9) to write:

$$dl \cong dl_x\left(1 + \frac{1}{2}\left(\frac{dy}{dx}\right)^2\right) \qquad (5.10)$$

Since we are only concerned with extension in the y-direction: $dl_y = dl - dl_x$, we use equation (5.3) to find the potential energy pumped into the system during vertical extension of the string segment:

$$dV_y = F \cdot dl_y = \frac{1}{2}F \cdot dl_x\left(\frac{\partial y(x,t)}{\partial x}\right)^2 = \frac{1}{2}Fk^2A^2dl_x\cos^2(kx - \omega t) \qquad (5.11)$$

With the help of the relations: $v = f \times \lambda$ and $\omega = 2\pi f$ along with equation (5.6) we can write:

$$Fk^2 = F\left(\frac{2\pi}{\lambda}\right)^2 = F\left(\frac{\omega}{v}\right)^2 = \mu_l\omega^2 \qquad (5.12)$$

so that

$$dV_y = \frac{1}{2}\mu_l \cdot dl_x\omega^2A^2\cos^2(kx - \omega t) \qquad (5.13)$$

Comparing equations (5.8) and (5.13), kinetic and potential energy are equally distributed in matter for every value of x and t in the vertical sense. Equation (5.8) shows kinetic energy as a function of the time component of y-displacement. According to equation (5.11) potential energy is dependent on y-displacement relative to x. This is satisfying in the sense that we are commonly reminded potential energy is energy of position. However potential energy can also be expressed as a function of time by rearranging equation (5.2) to the form: $dx = v \cdot dt$ and substituting into equation (5.11):

$$dV_y = \frac{1}{2}F \cdot dl_x\frac{1}{v^2}\left(\frac{\partial y(x,t)}{\partial t}\right)^2 = \frac{1}{2}\mu_l \cdot dl_x\omega^2A^2\cos^2(kx - \omega t) \qquad (5.14)$$

Note this is the same result obtained in equation (5.13).

The total energy can be found by summing kinetic and potential contributions, but a more facile form of total energy is its average value. For each displacement cycle (wavelength) or each time cycle (period), the average value of $\cos^2$ is ½. This gives an average total energy of:

$$dE_y = dT_y + dV_y = \frac{1}{2}\mu_l \cdot dl_x \omega^2 A^2 \tag{5.15}$$

It is also possible to define an average energy density as the energy per string segment:

$$\bar{E}_y = \frac{1}{2}\mu_l \omega^2 A^2 \tag{5.16}$$

Equation (5.16) is reminiscent of the total energy expression found for harmonic vibration in chapter 3. The connection can be further elucidated by noting from equation (5.3) that matter in the string is undergoing a vertical acceleration of:

$$a = \frac{\partial^2 y(x,\ t)}{\partial t^2} = -\omega^2 y(x,t) \tag{5.17}$$

Equation (5.17) is very reminiscent to the equation of motion for harmonic vibration in chapter 3. When compared to harmonic rotation, equation (5.17) is exactly of the same form as centripetal acceleration (equation (4.18)). Each particle in the string is undergoing simple harmonic motion as the transverse wave passes through. According to equation (5.17), string matter is being accelerated in proportion to its displacement, but in the opposite direction. However, equation (5.16) represents the averaged total energy density over a completed cycle. Unlike the harmonic vibrator, a translating string wave is *not* a conservative system where $dT + dV$ is the *same* dE for every time infinitesimal. A vibrating string does not exhibit alternating opposite increasing and decreasing magnitudes for its kinetic and potential energy distributions. To see this, consider vertical displacement of the string segment at its turning points. At these antinodes matter has zero velocity so $dT_y = 0$. At the same time its instantaneous rate of change of y with respect to x is also zero (zero tangential slope), so according to equation (5.11) it has $dV_y = 0$ as well. When the string segment passes through its equilibrium point, or at a node, it is at both maximum velocity and y slope with respect to x, so has simultaneously maximum values of dT_y and dV_y. Non-conservative behavior is possible because a cycle of oscillation is not an isolated system, they each pump energy in and out of one another, fueled by the mechanical oscillator at the end of the string.

Let us now change the dynamics of the propagating wave train. Instead of allowing it to travel eternally in the positive x-direction, suppose it encounters an attachment point to a wall, so that the string is of fixed length L. The point is ideally rigid, which causes the disturbance to reflect. As the wave cycle hits the barrier, it now propagates back in the negative x-direction. For the reflected wave:

$$y(x,t) = -A \sin \frac{2\pi}{\lambda}(x + vt) \tag{5.18}$$

Notice for the term $x + vt$ to remain constant requires negative values of x. A few restrictions are assumed to be consequences of the reflection point. First it represents a nodal point relative to the wave displacement. Secondly the wave experiences phase reversal, or reflects with 180° change relative to the nodal point. This is the reason that a negative sign is included in equation (5.18). The polarity change is a consequence of Newton's Third Law. The string pulling up on the wall at the reflection point causes a wall response of equal and opposite downward pull. Finally, for simplicity we assume the situation is devoid of damping factors, so not only does the disturbance propagate down the string with no loss of energy, but also there are no frictional losses at the reflection point.

Reflected waves must have wavelengths that match with waves travelling in the positive x sense, so that the nodal points of the two align and avoid destructive interference. If both ends of the string are restricted to be reflecting nodal points, resonance between waves of opposite polarity but equal wavelength can be sustained. This creates what is known as a *standing wave*. Figure 5.2 shows the first four possible resonating standing waves between two fixed points separated by a distance L. The figure uses a solid curve to represent a wave travelling from left to right, and a dashed curve for its opposite-polarity counterpart travelling from right to left. When this phenomenon is observed in an actual string, the direction of propagation is not discernible, and the string simply looks like a sequence of stationary loops. To avoid destructive interference, only an integer number of loops is possible for a sustainable standing wave. This creates an *overtone series*, where a string of length L can have standing waves that fit the wavelength pattern:

$$\lambda_n = \frac{2L}{n} \quad n = 1, 2, 3, \ldots \tag{5.19}$$

Figure 5.2 shows the *fundamental* or *first harmonic*: λ_1 with wavelength twice the string length. It also displays the first three overtones known as the second (λ_2), third (λ_3) and fourth (λ_4) *harmonics*.

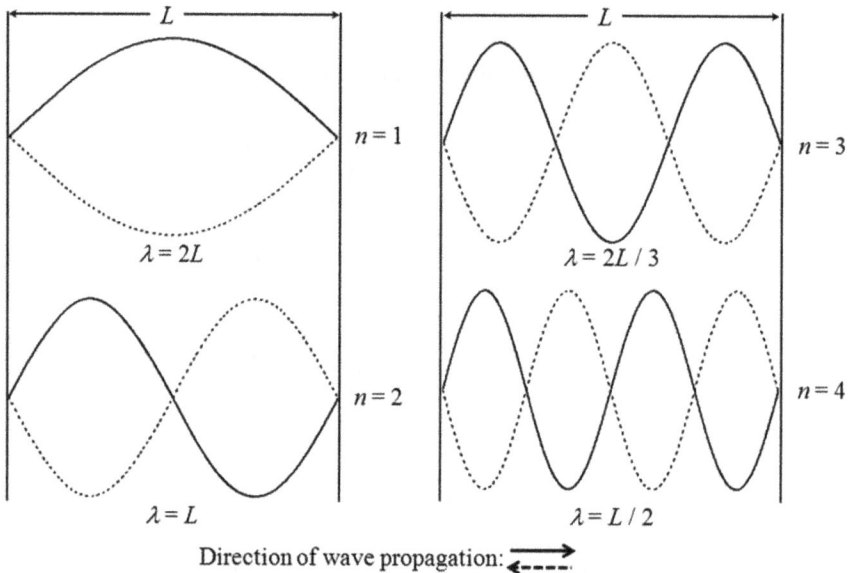

Figure 5.2. Wave patterns of the first four standing wave harmonics.

Using equation (5.6) and the relation: $v = f \times \lambda$ leads to an expression for the frequency or pitch of a vibrating string:

$$f_n = \frac{n}{2L}\sqrt{\frac{F}{\mu_l}} \tag{5.20}$$

Equation (5.20) was derived by Marsenne, the 17th century physicist and music theorist, as he investigated the behavior of stringed instruments. The driving force behind string oscillation can be plucking as for a guitar, hammering in a piano, or bowing as a violin. Of course the first two have several damping effects which cause the tone to eventually die out, but the third is a driven sustainable oscillator. Actual frequencies from stringed instruments are complex mixtures of many simultaneous overtones resonating down the string. Equation (5.20) shows that low pitch sounds are achieved by long, dense strings under little tension, with opposite parameters generating high pitch sounds. A piano with a range of over seven full octaves, requires a 2^7 increase in frequency, so not only is the length of its strings varied, but also the linear density and tension as well. To accomplish this solely from string length would require strings from 0.5 to 76 feet.

Although this discussion involved only transverse wave motion, a similar development can be applied to matter undergoing longitudinal wave motion such as sound. In this case, both the direction of propagation and matter disturbance occur in the same direction and matter is disrupted by rarefication and compression of air in pressure changes, rather than a travelling stretch.

5.2 de Broglie's analysis of translational motion

From simple observation of its pattern, the periodic nature of matter during vibration and rotation is very apparent. Macroscopic evidence of wavelike motion in translating matter is much too subtle to discern, still this supposition was made by de Broglie about a decade after the Bohr atomic model. The hypothesis of de Broglie arises from the *energy-momentum relation* of special relativity:

$$E^2 = (m_0 c^2)^2 + (pc)^2 \tag{5.21}$$

A photon has zero rest mass ($m_0 = 0$) and possesses energy: $E = h\nu$ (as seen in chapter 4, equation (4.33)). In the case of electromagnetic radiation we employ the more conventional representation of frequency, which uses the Greek letter nu: ν. With these facts, the energy-momentum relation for a photon becomes:

$$p = \frac{h\nu}{c} \tag{5.22}$$

Inserting the frequency-wavelength relation: $c = \lambda \times \nu$ into equation (5.22) gives an expression relating a photon's momentum to its wavelength:

$$p = \frac{h}{\lambda} \tag{5.23}$$

In his 1924 PhD thesis, de Broglie conjectured equation (5.23) not only applies to photons, but to matter as well. Legend has it that his graduate committee could not fully interpret the work, and suggested it could be more critically assessed by Einstein. When sent off to him Einstein responded in very short order, declaring that de Broglie had 'unraveled one of the secrets of the Universe'. Einstein's affirmation was all that the scientific community needed. Decades of ingenious theoretical work and painstaking experimental effort, all intended to settle the debate over the nature of matter—wave interaction, was now contained in a single encompassing equation of elegant simplicity. The prevailing philosophy became known as *wave–particle duality*, a conceptual marriage of phenomena of previously disparate origin.

It is easy to see how deBrogie's conjecture remained unbeknownst for centuries, in particular with regard to everyday observation. For instance a baseball (weight approx. 0.15 kg) travelling at 100 MPH has a de Broglie wavelength: $\lambda = 1.1 \times 10^{-34}$ m, which is certainly not going to throw off a batter. However, a 25 eV electron moves at 1/100th the speed of light and has: $\lambda = 2.4$ Å, which is proportional to the spacing between layers of an ionic crystal lattice. In 1913 Bragg and Bragg demonstrated diffraction of x-rays, electromagnetic radiation with wavelength on the order of magnitude as the spacing between crystal layers. When an x-ray beam was directed upon an ionic crystal acting as a diffraction grating, it generated an interference pattern. Between 1923 and 1927 Davisson and Germer performed experiments in which x-rays were replaced by accelerated electrons passing through an ionic crystal diffraction grating. The experiment generated an interference pattern, corroborating de Broglie suggestion and the wave-like character of matter. This historic experiment was conducted essentially 100 years after Young obtained an interference pattern in his double slit experiment, which had conclusively verified the wave-like character of light and Huygen's Principle.

To see what the de Broglie relation says about translation, consider matter of mass m confined to a 1-dimensional box of length L. It then possesses kinetic energy: $T = p^2/2m$. If it behaves with wave character the matter establishes a resonance pattern in the box. It is assumed that only wavelengths described by equation (5.19) are allowed, in order to reflect off the walls of the box without destructive interference. Using equations (5.23) and (5.19) we find:

$$E_n = \frac{p_n^2}{2m} = \frac{h^2}{2m\lambda_n^2} = \frac{h^2}{2m(2L/n)^2} = \frac{h^2 n^2}{8mL^2} \quad n = 1, 2, 3, \ldots \quad (5.24)$$

Equation (5.24) is one of the most celebrated results of quantum mechanics. It will be derived in chapter 6 from a completely different approach for a particle moving in a 1-dimensional box. In chapter 4, we saw how Bohr's quantization of angular momentum led to discrete energy levels in the one-electron atom. Here quantization of energy arises from application of a discrete linear momentum, due to allowed half integer wavelengths fitting the box dimension. The restrictions on wavelength create energy *levels* for the wavelike particle instead of the expected continuous energy distribution.

As an addition confirmation of the de Broglie relation, consider rotation of an electron around a proton. We assume it has a ground state ($N = 1$) wavelength equal to the circumference of its orbit. Using chapter 4, equation (4.27) the hydrogen atom has radius a_0, we allow a wavelength for the hydrogen atom electron which fits the circumference of its orbit according to:

$$\lambda = 2\pi \, a_0 \tag{5.25}$$

Using this in de Broglie's relation, equation (5.23), we find a linear momentum of:

$$p = \frac{\hbar}{a_0} \tag{5.26}$$

The kinetic energy of a hydrogen atom electron obeying the de Broglie relation is then:

$$T = \frac{p^2}{2\mu} = \frac{1}{2}\frac{\hbar^2}{\mu a_0^2} = \frac{1}{2}E_h \tag{5.27}$$

The total energy: $E_{tot} = T + V$. Since the atom has a potential of the form: $V(r) = cr^{-1}$, the virial theorem (see chapter 3, equation (3.15)) shows the kinetic and potential energy are related by: $2\langle T \rangle = -\langle V \rangle$. Combining both forms gives a total energy:

$$E_{tot} = T + V = -\frac{1}{2}\frac{\hbar^2}{\mu a_0^2} = -\frac{1}{2}E_h \tag{5.28}$$

which compared to chapter 4, equation (4.31) is the same as the kinetic energy of Bohr's ground state hydrogen atom electron.

What's the Matter with Waves?
An introduction to techniques and applications of quantum mechanics
William Parkinson

Chapter 6

Quantum translation

6.1 Stationary state wavefunctions

Alternative derivations of quantum effects exist, notably Heisenberg's matrix mechanics, Feynman's path integral formulation, and Dirac's generalized quantum electrodynamics. This work follows the development attributed to Schrödinger in 1925. The cornerstone of quantum mechanics is the *wavefunction* $\Psi(r, t)$, an eigenfunction to the quantum mechanical Hamiltonian. (For a discussion of eigenfunctions, see chapter 3. Hamiltonians are discussed in chapter 2.) In this case, the Hamiltonian is a linear partial differential equation known as the *time-dependent Schrödinger equation*:

$$i\hbar\frac{\partial\Psi(r, t)}{\partial t} = \hat{H}\Psi(r, t) \tag{6.1}$$

Equation (6.1) completely describes the time evolution of matter, whether in free motion or bound by some potential. A caret accent signifies the Hamiltonian is a mathematical *operator* to the wavefunction. It also contains imaginary factor $i = \sqrt{-1}$ and the quantum unit of angular momentum $\hbar$, which is Planck's constant divided by 2π introduced in chapter 4. The scale of $\hbar$ results in particularly marked effects on matter of dimension or mass at the atomic or molecular scale. More generalized forms of equation (6.1) are used in applications of other propagating waves such as Madelung's expression for hydrodynamics, non-linear optics and acoustics, and the Navier–Stokes equation of general fluid dynamics.

According to classical mechanics, the time evolution of matter is completely specified through knowledge of the system's initial conditions, and applying Hamilton's equations of motion (or equivalently by applying Newton's Second Law: $F = ma$). In quantum mechanics, the wavefunction describes the probabilistic fate of matter. From it can be extracted the most statistically likely outcome to physical stimuli but, unlike classical mechanics, the act of measuring one particular property can possibly have an irrevocable affect on the value of a subsequent

doi:10.1088/978-1-6817-4577-0ch6

property. Schrödinger's equation not only describes a system's time evolution, but also defines a valid wavefunction, because its mathematical construct is subject to the Hamiltonian describing a physical situation.

Mathematics requires Ψ is an eigenfunction of $\hat{H}$ that depends upon $\hat{V}$, the potential term which is part of $\hat{H} = \hat{T} + \hat{V}$. As in the classical case the quantum Hamiltonian has a component describing kinetic energy, signified as operator: $\hat{T}$. Wavefunctions must be simultaneous eigenfunctions of both, but the potential is the term which varies from physical situation to situation. In fact, the discrete spatial and angular momentum distributions that uniquely characterize quantization are inherent constructs of well-behaved solutions to eigenvalue problems containing a particular potential form.

Before delving into specific structures for the kinetic and potential energy operators, assume for the moment that neither is time dependent: $\hat{H}(r, t) = \hat{H}(r)$ only. A time-independent potential $\hat{V}(r)$ implies matter can only be influenced by static electric or magnetic fields, so is not subjected to electromagnetic radiation. If this is the case, the wavefunction can be expressed in product form:

$$\Psi(r, t) = \psi(r)\Phi(t) \tag{6.2}$$

and inserted into equation (6.1):

$$i\hbar\psi(r)\frac{\partial\Phi(t)}{\partial t} = \Phi(t)\hat{H}(r)\psi(r) \tag{6.3}$$

Dividing both sides of equation (6.3) by the product wavefunction places the differential equation into a form that has one side containing only t dependence, while the other exhibits only r dependence. In this fashion, a *separation of variables* has been accomplished:

$$i\hbar\frac{1}{\Phi(t)}\frac{\partial\Phi(t)}{\partial t} = \frac{1}{\psi(r)}\hat{H}(r)\psi(r) \tag{6.4}$$

Since the two sides of equation (6.4) are equal, but are individually functions of independent variables, each must be equal to a constant value which is both r and t independent. Dimensional analysis shows this constant to have units of energy. Setting the left-hand side of equation (6.4) equal to a constant specified as E then rearranging leads to the eigenvalue equation:

$$\frac{\partial\Phi(t)}{\partial t} = \frac{E}{i\hbar}\Phi(t) \tag{6.5}$$

Eigenfunctions of equation (6.5) have the form:

$$\Phi(t) = Ne^{Et/i\hbar} \tag{6.6}$$

where N is a time-independent constant. The right-hand side of equation (6.4) is set equal to the same constant and rearranges to the familiar form known as the *time-independent Schrödinger equation*:

$$\hat{H}(r)\psi(r) = E\psi(r) \tag{6.7}$$

Equation (6.7) is an eigenvalue equation with operator: $\hat{H}$, eigenfunction: $\psi(r)$, and eigenvalue: E. The $\psi(r)$ are eigenfunctions of a time-independent Hamiltonian, valid when $\hat{V}(r, t) = \hat{V}(r)$ only. The resulting time-independent states are referred to as *stationary states*. The remainder of this book will focus on stationary state quantum mechanics and properties of solutions to the time-independent Schrödinger equation. Each eigenfunction to the stationary state eigenvalue problem possesses a single, definite energy. Most often there will be a collection of valid eigenfunctions to the same problem which collectively represent a discrete, or *quantized*, set of energy levels—one for each of the independent stationary state solutions to a given Hamiltonian. Transitions between stationary state energy levels, in either emission or absorption processes, require matter to emit or absorb photons of the correct frequency according to the Bohr condition: $\Delta E = h\nu$. Coupling stationary states $\psi(r) \rightarrow \psi'(r)$ via electromagnetic radiation introduces time dependence to the potential $\hat{V}(r, t)$, resulting in a superposition of time-dependent energy eigenstates $\Psi(r, t)$ allowing the transition.

Valid wavefunctions are eigenstates of the Hamiltonian describing a particular physical situation. There are additional mathematical requirements ψ must obey that were first stipulated by Born. Over all space ψ must be: (i) single valued, (ii) finite, (iii) continuous, and (iv) have a continuous first derivative. Examples of valid and invalid wavefunctions are presented in figure 6.1.

6.2 Unconstrained one-dimensional translation

We will take an opposite tack from the classical story and begin by applying the Schrödinger equation to the modes of motion matter undertakes while translating. First consider the quantum mechanical motion of a *free particle*, one with motion under the influence of no potential. We will not concern ourselves with how it got into motion, we will simply accept that it will remain in that motion, according to Newton, and translate in a straight trajectory. The coordinate system is aligned so that motion is entirely along the x-axis. Since $\hat{V} = 0$ the Hamiltonian is therefore:

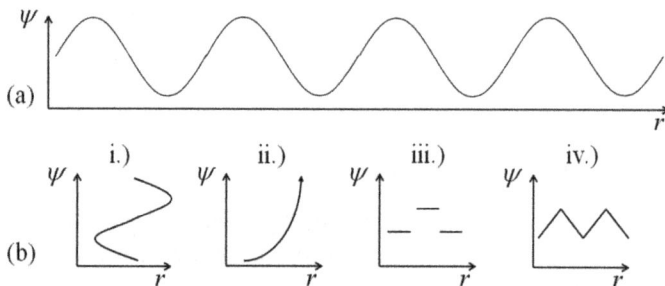

Figure 6.1. (a) A wavefunction satisfying all of Born's mathematical conditions. (b) Examples of wavefunctions which violate criteria (i)–(iv).

$\hat{H} = \hat{T}$. Using the kinetic energy operator shown in table 6.1 the Schrödinger equation is:

$$-\frac{\hbar^2}{2m}\frac{d^2}{dx^2}\psi = E\psi \tag{6.8}$$

This is an eigenvalue equation very similar in form to the classical harmonic oscillator problem discussed in chapter 3 (see equation (3.7)). As done there, valid solutions come in a general imaginary form:

$$\psi_1 = Ae^{+ikx} + Be^{-ikx} \tag{6.9}$$

as well as a real expression:

$$\psi_2 = C \sin kx + D \cos kx \tag{6.10}$$

It should be verified by direct substitution into equation (6.8) that either solution is an eigenfunction of the free particle Hamiltonian, with the definition of k (SI unit: m^{-1}):

$$k = \sqrt{\frac{2mE}{\hbar^2}} \tag{6.11}$$

Although both ψ_1 and ψ_2 are mathematically valid energy eigenfunctions, the typically preferred solution is an appropriate representative of free particle momentum. Applying the momentum operator: $\hat{p}_x = \frac{\hbar}{i}\frac{d}{dx}$ to equation (6.10) gives:

$$\hat{p}_x\psi_2 = \frac{k\hbar}{i}(C \cos kx - D \sin kx) \tag{6.12}$$

This shows a trigonometric solution is not a proper momentum eigenfunction. The exponential solution is indeterminate to a definite momentum state because equation (6.9) is a *superposition of states*, simultaneously representing translation in *both* the positive and negative x-directions. Imposing a *boundary condition* that the particle is moving +x *collapses* ψ_1 to the specific case where coefficient A is some (possibly complex) finite value and coefficient $B = 0$. The particle then has momentum:

$$\hat{p}_x\psi_1 = +\frac{\hbar i k}{i}Ae^{+ikx} = +\hbar k\psi_1 \tag{6.13}$$

If it instead is translating in the negative x sense, then $A = 0$, $B \neq 0$, and the momentum eigenvalue is: $-\hbar k$.

Using the relation: $e^{\pm i\theta} = \cos(\theta) \pm i \sin(\theta)$, the exponential wavefunction: $\psi_1 = Ae^{+ikx}$ is a complex *plane wave* of amplitude A propagating in the positive x-direction. The real part is a cosine function trailed by an imaginary sine

Table 6.1. Quantum Mechanical Operators.

Quantity	General form	x-component	y-component	z-component
Position	$\hat{r} = r\times$	$\hat{x} = x\times$	$\hat{y} = y\times$	$\hat{z} = z\times$
Momentum	$\hat{p} = \dfrac{\hbar}{i}\dfrac{\partial}{\partial r} = \dfrac{\hbar}{i}\nabla$	$\hat{p}_x = \dfrac{\hbar}{i}\dfrac{\partial}{\partial x}$	$\hat{p}_y = \dfrac{\hbar}{i}\dfrac{\partial}{\partial y}$	$\hat{p}_z = \dfrac{\hbar}{i}\dfrac{\partial}{\partial z}$
Kinetic energy	$\hat{T} = \dfrac{\hat{p}^2}{2m} = -\dfrac{\hbar^2}{2m}\nabla^2$	$\hat{T}_x = -\dfrac{\hbar^2}{2m}\dfrac{\partial^2}{\partial x^2}$	$\hat{T}_y = -\dfrac{\hbar^2}{2m}\dfrac{\partial^2}{\partial y^2}$	$\hat{T}_z = -\dfrac{\hbar^2}{2m}\dfrac{\partial^2}{\partial z^2}$
Angular momentum	$\hat{L} = \dfrac{\hbar}{i}\nabla\times r$	$\hat{L}_x = \dfrac{\hbar}{i}\left(y\dfrac{\partial}{\partial z} - z\dfrac{\partial}{\partial y}\right)$	$\hat{L}_y = \dfrac{\hbar}{i}\left(z\dfrac{\partial}{\partial x} - x\dfrac{\partial}{\partial z}\right)$	$\hat{L}_z = \dfrac{\hbar}{i}\left(x\dfrac{\partial}{\partial y} - y\dfrac{\partial}{\partial x}\right)$

component propagating with 90° phase difference. Either through rearrangement of equation (6.11) or by solving the energy eigenvalue problem we find:

$$\hat{T}\psi_1 = -\frac{\hbar^2}{2m}\frac{d^2}{dx^2}Ae^{+ikx} = \frac{\hbar^2 k^2}{2m}\psi_1 = E\psi_1 \tag{6.14}$$

As to be expected, energy and momentum are related by: $E = p_x^2/2m$. More importantly, the only restrictions on k are that it is (a) positive and (b) real. Assertion (a) is requisite for the particle to be translating in the positive x sense. Assertion (b) is a consequence of both energy and momentum being measurable quantities, so must therefore be represented by real numbers. There is no limitation on the magnitude of k, meaning the energy and momentum of a quantum mechanical free particle are both continuous quantities that in theory can take any numeric value.

6.3 One-dimensional translation in a box

So far no restrictions are placed on translation of the quantum mechanical free particle. In other words it has domain: $-\infty \leqslant x \leqslant +\infty$. Suppose that instead of free range it encounters infinite barriers at each end, which confines its motion to a fixed region of space we will specify to be: $0 \leqslant x \leqslant L$. Between these boundaries, it remains independent of any potential. These are the conditions constraining a *particle in a 1-dimensional box*. Within the box, the Hamiltonian is that of a free particle, and the time independent Schrödinger equation is the same form as equation (6.8). As a consequence equations (6.9) or (6.10) represent valid solutions within the confines of the box, as well as the form in equation (6.11) for the value k.

Things get interesting when the superposition of states represented in the general solutions are subjected to the boundary conditions of the box. Constraining the particle to remain in the box at its left edge requires the wavefunction vanish at $x = 0$. From equation (6.9) the boundary condition: $\psi_1(0) = 0$ is only fulfilled if $A = B = 0$, eliminating the exponential form as a viable eigenfunction. From equation (6.10) we see that $\psi_2(0) = 0$ requires: $D = 0$. We have collapsed the real trigonometric particle in a box solution to: $\psi = C \sin kx$. Confining the particle to the box at its right edge further requires the wavefunction vanish at $x = L$. This can be satisfied by setting $C = 0$, which does not accomplish anything but trivialize the solution. Instead the boundary condition: $\psi(L) = 0$ is fulfilled for a sine function creating a node at the right box edge, which occurs whenever the product kL is equal to integer values of π. As a result there are an infinite number of solutions, that satisfy the condition: $k = n\pi/L$. The particle in a box wavefunctions have general form:

$$\psi_n = C \sin\left(\frac{n\pi x}{L}\right) \quad n = 1, 2, 3, \dots \tag{6.15}$$

Figure 6.2 displays plots of the first four particle in a box wavefunctions. Note the boundary conditions create a zero amplitude nodal point at each barrier. The wavefunctions in figure 6.2 bear a striking resemblance to the functions we

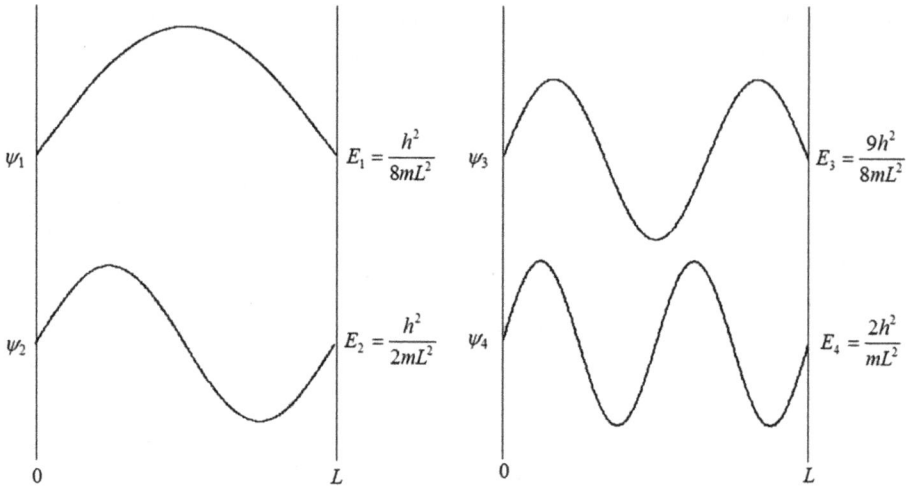

Figure 6.2. Plots of the first four 1-dimensional particle in a box wavefunctions.

encountered in chapter 5, the standing waves that are formed by an oscillating string that reflects between two infinite barriers. You should refer to figure 5.2, and note these classical waveforms have exactly the same structure as their quantum mechanical counterparts. The main feature of either the classical or quantum waveforms is a sequential increase in the number of nodes corresponding to the integer value of n. In both cases, this simple integer is the central characterizing feature of the solutions. As further systems are investigated, it will quickly become apparent that, in spite of the complexity of the mathematics, quantum mechanical wavefunctions are invariably subject to and dependent upon straightforward elementary numbers. These values, which play basic roles in the energetics and structure of quantum mechanical solutions, are known as *quantum numbers*.

Surprisingly, the particle in a box solutions are not eigenfunctions of either position or momentum. This appears irreconcilable with the postulates of quantum mechanics, which mandate the wavefunction possess all physical information describing a system. It will take techniques introduced in the next chapter to see how properties are extracted from wavefunctions.

Since the expressions in equation (6.15) solve the time independent Schrödinger equation for this system, they are energy eigenfunctions. As was done for the free particle, the energy of a particle in a box can thus be found by either substituting the condition on k into equation (6.11), or through the eigenvalue problem:

$$\hat{T}\psi_n = -\frac{\hbar^2}{2m}\frac{d^2}{dx^2}C\sin\left(\frac{n\pi x}{L}\right) = \frac{\hbar^2 n^2 \pi^2}{2mL^2}\psi_n = E_n\psi_n \quad n = 1, 2, 3, \dots \quad (6.16)$$

Simplifying equation (6.16) shows that the boundary conditions imposed on a free particle confined to a restricted box of length L creates a discrete energy distribution with energy levels given by:

$$E_n = \frac{h^2 n^2}{8mL^2} \quad n = 1, 2, 3, \ldots \qquad (6.17)$$

Values for the first four energies are presented with the wavefunctions in figure 6.2.

It is noted that both the nodal structure and particle energy increases with increasing quantum number n. Referring back to the previous chapter's discussion of translation, we there introduced a standing wave constraint (see chapter 5, equation (5.19)) into the de Broglie relation (see chapter 5, equation (5.23)). The energy determined in that case (see chapter 5, equation (5.24)) is exactly the same energy expression as equation (6.17). In chapter 5 the energy was obtained by treating the resonant wavelength of translation via the de Broglie relation as opposed to this case where it is obtained via the Schrödinger equation. It is satisfying to see that both approaches lead to the same conclusion.

PARALLEL INVESTIGATION: Verify that the wavefunctions: $\chi_n = C \cos\left(\frac{n\pi x}{2L}\right)$, $n = 1, 3, 5, \ldots$ are eigenfunctions to the particle in a box Hamiltonian under the boundary conditions: $V(x) = \begin{cases} \infty & x = -L \\ 0 - L < x < L \\ \infty & x = L \end{cases}$ with eigenvalues of: $E_n = \frac{h^2 n^2}{32mL^2}$ $n = 1, 3, 5, \ldots$

6.4 Multi-dimensional translation in a box

Instead of a particle free to translate in a 1-dimension space, suppose it is in an actual 2-dimensional box of length L_x and width L_y. Again the assumption is that no potential acts on the particle within the box, but its edges form a confining, inescapable barrier. Inside the box where the particle is potential-free, the eigenvalue problem takes the form:

$$\hat{H}(x, y)\psi(x, y) = -\frac{\hbar^2}{2m}\left(\frac{\partial^2}{\partial x^2} + \frac{\partial^2}{\partial y^2}\right)\psi(x, y) = E\psi(x, y) \qquad (6.18)$$

Any general motion in the 2-dimensional space of the box can be expressed as a combination of independent motions along the Cartesian x and y directions. In other words the Hamiltonian describing translation can be separated into individual terms each dependent on a single variable:

$$\hat{H}(x, y) = \hat{H}(x) + \hat{H}(y) = -\frac{\hbar^2}{2m}\frac{\partial^2}{\partial x^2} - \frac{\hbar^2}{2m}\frac{\partial^2}{\partial x^2} \qquad (6.19)$$

The solution to a separable Hamiltonian is a separable eigenfunction expressed as a product of functions, each dependent on a single variable:

$$\psi(x, y) = C\varphi(x)\varphi(y) \qquad (6.20)$$

This assertion is apparent when equation (6.18), (6.19), and (6.20) are combined:

$$C\varphi(y)\frac{\partial^2}{\partial x^2}\varphi(x) + C\varphi(x)\frac{\partial^2}{\partial y^2}\varphi(y) = -\frac{2mE}{\hbar^2}C\varphi(x)\varphi(y) \tag{6.21}$$

Dividing equation (6.21) by equation (6.20) completes the separation of variables:

$$\frac{1}{\varphi(x)}\frac{\partial^2}{\partial x^2}\varphi(x) + \frac{1}{\varphi(y)}\frac{\partial^2}{\partial y^2}\varphi(y) = -\frac{2mE}{\hbar^2} \tag{6.22}$$

In equation (6.22), the right-hand side is a function of constants only. Since each term on the left-hand side is an independent function of a separate variable, they must each be equal to their own constants that are labeled E_x and E_y, subject to the condition: $E = E_x + E_y$. The result is two independent equations:

$$\begin{aligned}\frac{1}{\varphi(x)}\frac{\partial^2}{\partial x^2}\varphi(x) &= -\frac{2mE_x}{\hbar^2} \\ \frac{1}{\varphi(y)}\frac{\partial^2}{\partial y^2}\varphi(y) &= -\frac{2mE_y}{\hbar^2}\end{aligned} \tag{6.23}$$

Multiplying the top equation through by: $\varphi(x)$ and the bottom equation through by: $\varphi(y)$ produces two independent particle in a box eigenvalue problems, subject to the constraints: $0 \leqslant x \leqslant L_x$ and $0 \leqslant y \leqslant L_y$, respectively. Boundary value problems in this same form have already been discussed in section 6.2. Each has a solution of the type presented in equation (6.15). Application of the two boundary conditions $\varphi(x = L_x) = 0$ or $\varphi(y = L_y) = 0$ again restrict the solutions to integer multiples of π, but require independent quantum numbers: n_x and n_y. The 2-dimensional particle in a box wavefunction is then:

$$\psi_{n_x,n_y}(x, y) = C \sin\left(\frac{n_x \pi x}{L_x}\right)\sin\left(\frac{n_y \pi y}{L_y}\right) \quad n_x = 1, 2, 3, \dots \quad n_y = 1, 2, 3, \dots \tag{6.24}$$

The total energy is the sum of energies from the independent eigenvalue problems:

$$E_{n_x,n_y} = \frac{h^2}{8m}\left(\frac{n_x^2}{L_x^2} + \frac{n_y^2}{L_y^2}\right) \quad n_x = 1, 2, 3, \dots \quad n_y = 1, 2, 3, \dots \tag{6.25}$$

The energy representation is discrete, but now is a function of two quantum numbers n_x and n_y. An interesting consequence results when the 2-dimensional space is a square rather than a rectangle. Equation (6.25) shows that any two eigenstates of the form: $n_x \neq n_y$ have energy eigenvalues: $E_{ij} = E_{ji}$. These are known as *degenerate energy levels*.

Figure 6.3 plots the first four particle in a 2-dimensional box wavefunctions for the special case that: $L_x = L_y = L$. The perspective is such that you are looking down on the box, with the wavefunction displayed via a contour plot of regions with equal wave amplitude. Each antinode is indicated as being a 2-dimensional wave peak or trough according to the sign at its vertex. The spacing between contour lines is an indicator of the steepness of the wavefunction in that region. In figure 6.2, a

1-dimensional particle in a box wavefunction ψ_n shows $n - 1$ nodal points of zero wave amplitude. The 2-dimensional functions in figure 6.3 possess nodal *planes*, which number n_x-1 in the x-direction and n_y-1 in the y-direction. Another notable feature in figure 6.3 is the spatial relationship demonstrated by degenerate wavefunctions: ψ_{21} and ψ_{12}. Rotating one by a 90° symmetry operation converts it into the other. Degenerate energy levels are common occurrences across quantum mechanics. Invariably, symmetry properties of wavefunctions and their energetic degeneracy go hand in hand.

Separation of variables can be extended to free translation in a 3-dimensional space in a straightforward fashion. Inside the region there is no potential, so the system has Hamiltonian:

$$\hat{H}(x, y) = \hat{H}(x) + \hat{H}(y) + \hat{H}(z) = -\frac{\hbar^2}{2m}\nabla^2 = -\frac{\hbar^2}{2m}\left(\frac{\partial^2}{\partial x^2} + \frac{\partial^2}{\partial y^2} + \frac{\partial^2}{\partial z^2}\right) \quad (6.26)$$

for a particle confined to the coordinates: $0 \leqslant x \leqslant L_x$, $0 \leqslant y \leqslant L_y$, and $0 \leqslant z \leqslant L_z$. The Cartesian variables are separable, allowing a product wavefunction consisting of three independent components. Applying boundary conditions that ψ must vanish at the cube edges leads to the particle in a cube wavefunction:

$$\psi_{n_x,n_y,n_z}(x, y, z) = C \sin\left(\frac{n_x\pi x}{L_x}\right)\sin\left(\frac{n_y\pi y}{L_y}\right)\sin\left(\frac{n_z\pi z}{L_z}\right) \quad (6.27)$$

$n_x = 1, 2, 3, \ldots \; n_y = 1, 2, 3, \ldots \; n_z = 1, 2, 3, \ldots$

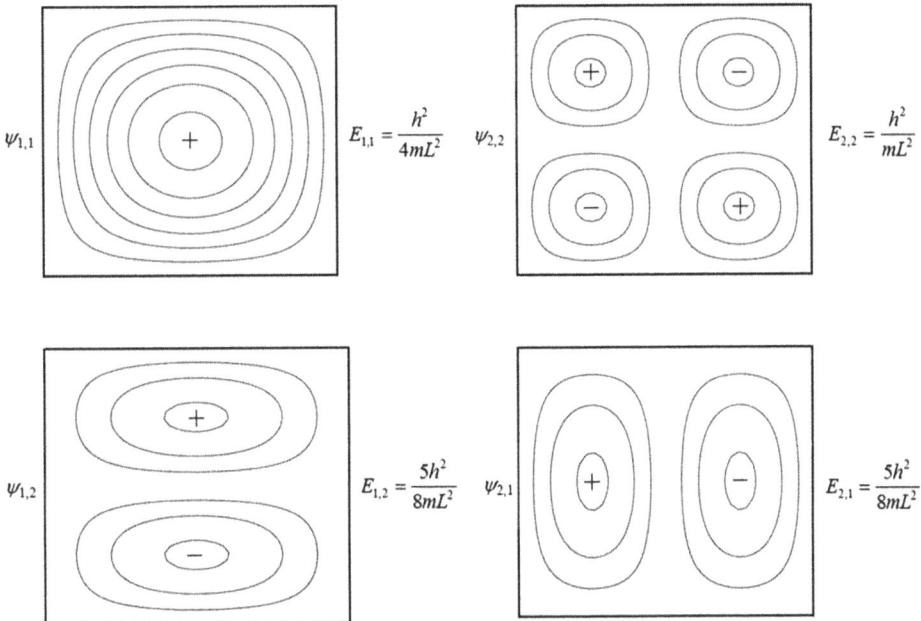

Figure 6.3. Plots of 2-dimensional particle in a box wavefunctions.

with energy eigenvalues:

$$E_{n_x,n_y,n_z} = \frac{h^2}{8m}\left(\frac{n_x^2}{L_x^2} + \frac{n_y^2}{L_y^2} + \frac{n_z^2}{L_z^2}\right) \quad n_x = 1, 2, 3, \ldots \quad n_y = 1, 2, 3, \ldots$$

$$n_z = 1, 2, 3, \ldots$$

(6.28)

Notice a cubic system with $L_x = L_y = L_z$ presents a multitude of opportunities for degenerate energy levels.

PARALLEL INVESTIGATION: Verify that the wavefunctions:

$\chi_{n_x,n_y,n_z} = C \cos\left(\frac{n_x \pi x}{2L_x}\right) \cos\left(\frac{n_y \pi y}{2L_y}\right) \cos\left(\frac{n_z \pi z}{2L_z}\right)$, are eigenfunctions to the particle in a

cube Hamiltonian under the boundary conditions: $V(x) = \begin{cases} \infty & x = -L \\ 0 & -L < x < L \\ \infty & x = L \end{cases}$ and similar

forms for $V(y)$ and $V(z)$. Show that the eigenvalues are: $E_n = \frac{h^2}{32m}\left(\frac{n_x^2}{L_x^2} + \frac{n_y^2}{L_y^2} + \frac{n_z^2}{L_z^2}\right)$

with quantum numbers of odd integers: $n_x = 1, 3, 5, \ldots$ $n_y = 1, 3, 5, \ldots$ $n_z = 1, 3, 5, \ldots$

Before moving on to see what quantum theory has to say about rotation and vibration, let us take advantage of the relatively straightforward results presented in this section. We now have a model for matter completely free to move within a container, but from which it has no possibility of escape. The solutions turn out to have a familiar sinusoidal form inherent to wave motion, in many ways similar to the classical problem of standing waves on a vibrating string. In both cases, the waves 'fit' the box with a requirement of nodal reflections at the barrier. The properties of well-behaved quantum mechanical solutions require a discrete rather than continuous energy distribution. In the next chapter we continue our investigation by applying the model to demonstrate some of the other important features, techniques and oddities of quantum mechanics.

IOP Concise Physics

What's the Matter with Waves?
An introduction to techniques and applications of quantum mechanics
William Parkinson

Chapter 7

Interpreting quantum mechanics

7.1 The probability density

The wavefunction itself may in general be real, imaginary or a complex function of both real and imaginary components. Of course physical properties must be real, so a recipe is postulated in quantum mechanics for determining such values from any wavefunction independent of its form. As mentioned previously, this process is statistical in nature, performed via distribution theory. An illustrative example of statistical averaging outcomes is supplied by the case of six students taking an exam, and earning the following scores: 50, 62, 78, 78, 84, 92. The mean test score is determined according to:

$$\langle \text{avg} \rangle = \frac{50 + 62 + 78 + 78 + 84 + 92}{6} \tag{7.1}$$

This can also be written:

$$\langle \text{avg} \rangle = \frac{1}{6}*50 + \frac{1}{6}*62 + \frac{1}{3}*78 + \frac{1}{6}*84 + \frac{1}{6}*92 \tag{7.2}$$

There is a probability of 1/6 (or a 1/6 chance) of obtaining the grade 50, 62, 84, or 92 and a probability of 1/3 (or 1/3 chance) of scoring 78. Another way to write this is:

$$\langle q \rangle = \sum_i P(q)_i q_i \tag{7.3}$$

Here the average value $\langle q \rangle$ is obtained by summing the product of score q_i with its weight or probability $P(q)_i$. The weighting factors fulfill the general requirement of unit probability:

$$\sum_i P(q)_i = 1 \tag{7.4}$$

doi:10.1088/978-1-6817-4577-0ch7

Relating the above to outcomes in quantum mechanics requires transition from a discrete to continuous distribution, or that summation becomes integration. In addition, a quantum mechanical probability function must be postulated. This was done by Born, who stated that the wavefunction had a *probability density* found from its modulus square over all space. The Born Law (or Born Rule) is:

$$|\psi(r)|^2 = \psi(r)^*\psi(r) = P(r) \tag{7.5}$$

where $\psi(r)^*$ is the wavefunction complex conjugate (found by changing the sign of any imaginary component). The modulus square of any value is a real positive number, no matter if the value itself is positive or negative, real, imaginary or complex. Therefore the probability density of any wavefunction is a real, positive quantity. Even for sinusoidal functions it is not only real but is also positive everywhere in space.

As a mental picture of probability density, think of smashing a particle into an infinite number of pieces which scatter throughout all space. In various regions fragments aggregate to differing degrees, some places with none at all, some a sparse distribution, while others amass a noticeably pronounced collection. Throughout space, the relative consistency of fragment density represents the likelihood of the particle being at that particular point. The particle is reconstituted only by gathering all its components, an action equivalent to integrating the probability density. The discrete probability constraint of equation (7.4) has thus been converted to continuous form:

$$\int |\psi(r)|^2 \, d\tau = \int \psi(r)^*\psi(r) \, d\tau = 1 \tag{7.6}$$

The volume element in equation (7.6) has the following 3-dimensional Cartesian or spherical polar forms:

$$\int d\tau = \int_{-\infty}^{+\infty} dx \int_{-\infty}^{+\infty} dy \int_{-\infty}^{+\infty} dz = \int_0^{\infty} r^2 \, dr \int_0^{\pi} \sin\theta \, d\theta \int_0^{2\pi} d\varphi \tag{7.7}$$

There would also be a time dimension for non-stationary states. In addition, a linearly independent non-classical spin coordinate must be included as well, an aspect which will be explored in chapter 11. If the wavefunction represents a multiple particle system there will be separate spin and spatial coordinates for each component.

The need for square integrable wavefunctions to determine properties in quantum mechanics has led to what is now referred to as Dirac notation. A so-called 'ket' is used to represent a wavefunction in the form: $|\psi\rangle$, and its complex conjugate is depicted as a 'bra':

$$|\psi\rangle^* = \langle\psi| \tag{7.8}$$

Symbolically attaching bra to ket mathematically implies an accompanying integration over all space. For example, expressing equation (7.6) in bra-ket notation:

$$\langle \psi | \psi \rangle = \int \psi^* \psi \, d\tau = 1 \tag{7.9}$$

Eigenfunctions to a particular Hamiltonian are very often adjusted to satisfy equation (7.9), which is known as the *normalization condition*. This is accomplished by multiplying an arbitrary wavefunction by a constant known as the normalization factor N:

$$|\psi\rangle = |N\varphi\rangle \tag{7.10}$$

Note that multiplying an eigenfunction by a constant does not alter its property of being an eigenfunction. In general the normalization factor can be imaginary or complex as well as real, so that N should technically be complex conjugated for a bra function. Applying the normalization condition equation (7.9) to equation (7.10) allows the normalization factor to be determined:

$$N = \frac{1}{|\langle \varphi | \varphi \rangle|^{1/2}} \tag{7.11}$$

The prefactor C included in the solution of the 1-, 2-, and 3-dimensional particle in a box wavefunctions found in chapter 6, equations (6.15), (6.24), and (6.26) can now be replaced by a normalization constant. Imposing equation (7.9) on the 1-dimensional particle in a box over the limits: $0 \leqslant x \leqslant L$ gives:

$$\langle \psi_n | \psi_n \rangle = 1 = C^2 \int_0^L \sin^2\left(\frac{n\pi x}{L}\right) dx = C^2\left(\frac{L}{2}\right) \tag{7.12}$$

The right-hand side of equation (7.12) should be verified using the table of integrals in appendix B. Note that the integral solution is independent of quantum number n, or the same normalization factor applies to all functions. The normalized particle in a box expressions are:

$$|\psi_n\rangle = \left(\frac{2}{L}\right)^{1/2} \sin\left(\frac{n\pi x}{L}\right) \tag{7.13}$$

PARALLEL INVESTIGATION: Verify the normalized form for particle in a 1-dimensional box wavefunctions have the form: $|\chi_n\rangle = \frac{1}{L^{1/2}} \cos\left(\frac{n\pi x}{2L}\right)$, $n = 1, 3, 5, \ldots$ over the range: $-L \leqslant x \leqslant L$.

Normalizing the 2-dimensional particle in a box requires a double integral over independent variables x and y. To accomplish this we insert equation (6.24) into the normalization condition from equation (7.11) and integrate over the limits of the separable area element: $dx \cdot dy$. The two individual integrals each are of the same form as equation (7.12):

$$\langle \psi_{n_x,n_y} | \psi_{n_x,n_y} \rangle = 1 = C^2 \int_0^{L_x} \sin^2\left(\frac{n_x \pi x}{L_x}\right) dx \int_0^{L_y} \sin^2\left(\frac{n_y \pi y}{L_y}\right) dy$$

$$= C^2\left(\frac{L_x}{2}\right)\left(\frac{L_y}{2}\right) \qquad (7.14)$$

This leads to normalized functions of the form:

$$|\psi_{n_x,n_y}\rangle = \frac{2}{A^{1/2}} \sin\left(\frac{n_x \pi x}{L_x}\right)\sin\left(\frac{n_y \pi y}{L_y}\right) \quad n_x = 1, 2, 3, \ldots \quad n_y = 1, 2, 3, \ldots \qquad (7.15)$$

where the box area is: $A = L_x \times L_y$.

7.2 Eigenvectors and basis sets

The complete set of wavefunctions in equations (7.13) or (7.15) constitute a 1-dimensional array known as an *eigenvector* of elements. In addition to fulfilling the normalization condition, any two elements of an eigenvector have an *inner product* which obeys:

$$\langle \psi_i | \psi_j \rangle = \delta_{ij} \qquad (7.16)$$

Equation (7.16) contains the *Kronecker delta*:

$$\delta_{ij} = \begin{array}{ll} 0 & \text{if: } i \neq j \\ 1 & \text{if: } i = j \end{array} \qquad (7.17)$$

The bottom condition says that individual eigenvector elements are normalized, or exhibit unit probability when integrated over all space. The top condition states any two elements are *orthogonal* to one another. Collectively, all elements of the eigenvector form an *orthonormal set*. The orthogonality condition can be demonstrated using any two elements $i \neq j$ of the 1-dimensional particle in a box eigenvector, for instance:

$$\langle \psi_3 | \psi_4 \rangle = 0 = \frac{2}{L} \int_0^L \sin\left(\frac{3\pi x}{L}\right) \cdot \sin\left(\frac{4\pi x}{L}\right) dx \qquad (7.18)$$

Referring to the list of standard integrals or using numeric software, it should be verified that the integral on the right hand side is zero.

PARALLEL INVESTIGATION: Verify that the $n = 1$ and $n = 3$ wavefunctions of the form: $|\chi_n\rangle = \frac{1}{L^{1/2}} \cos\left(\frac{n\pi x}{2L}\right)$, $n = 1, 3, 5, \ldots$ are orthogonal over the range: $-L \leqslant x \leqslant L$.

Orthogonal functions have zero net *overlap* with one another, meaning the complex modulus product of amplitudes of orthogonal functions has zero net coincidence when infinitesimally summed over all space. Equation (7.18) corroborates this statement mathematically, but it is instructive to investigate this concept graphically. Figure 7.1(a) gives amplitude plots of the eigenvector elements $|\psi_3\rangle$ and $|\psi_4\rangle$ over their domain: $0 \leqslant x \leqslant L$. The figure is enhanced by placing a line through the nodal points with labels indicating every quarter interval of the box. This plot clearly shows the amplitudes of each have many regions of coincidence leading to constructive interference of their wave forms. However in other areas the waves are in opposite phase and exhibit destructive interference. Classical superposition of these wave forms is performed a by adding amplitudes. For sound waves this would for instance produce an acoustical beat pattern. In quantum mechanics the quantity of interest is instead the amplitude product, which is displayed in figure 7.1(b). Integration sums area under all curves. It is noted the product function in figure 7.1

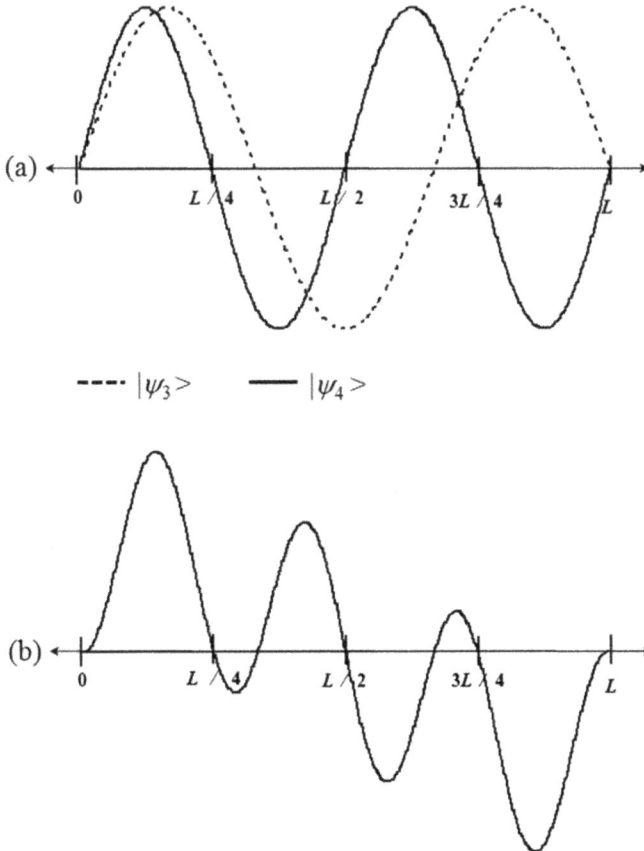

$$---- \quad |\psi_3\rangle \qquad \text{———} \quad |\psi_4\rangle$$

Figure 7.1. Plot of the amplitude of individual eigenvector elements: $|\psi_3\rangle = \left(\dfrac{2}{L}\right)^{1/2} \sin\left(\dfrac{3\pi x}{L}\right)$ and $|\psi_4\rangle = \left(\dfrac{2}{L}\right)^{1/2} \sin\left(\dfrac{4\pi x}{L}\right)$ over the range: $0 \leqslant x \leqslant L$. (b) Plot of the product amplitude $|\psi_4\rangle\langle\psi_3|$ over the range: $0 \leqslant x \leqslant L$.

(b) forms six distinct regions, three each being equal in area but opposite in sign. Summing the net area under all portions of the curves gives visual confirmation of zero net overlap between $|\psi_3\rangle$ and $|\psi_4\rangle$.

An arbitrary set of eigenfunctions can be used to construct an *overlap matrix* with elements:

$$S_{ij} = \langle \psi_i | \psi_j \rangle \tag{7.19}$$

According to equation (7.16), an orthonormal set of eigenfunctions have an overlap matrix equal to the unit matrix: $\mathbf{S} = \mathbf{1}$. The overlap matrix is Hermitian, or self-adjoint, meaning its elements obey the condition:

$$S_{ij} = S_{ji}^* \tag{7.20}$$

In the case of an orthonormal basis, Hermitian character of the overlap matrix is a trivial result. Even for a set of non-orthogonal complex functions the overlap matrix is often real, a result of complex conjugation of imaginary bra exponential components and their subsequent multiplication to ket exponentials of the opposite sign. However in general $\mathbf{S}$ can have complex elements.

Orthonormal eigenvector elements have no linear dependencies, and form a *basis* which *span* a vector space. An arbitrary function $|\Phi\rangle$ can be constructed from a linear combination, or superposition, of basis set elements $|\psi_j\rangle$:

$$|\Phi\rangle = N \sum_j c_j |\psi_j\rangle = \sum_j c_j' |\psi_j\rangle \tag{7.21}$$

The expansion coefficients c_j in general may be complex, and are chosen to fulfill the normality condition so that:

$$N = \frac{1}{|\langle \Phi | \Phi \rangle|^{1/2}} = 1 = \frac{1}{\left| \sum_k \sum_j c_j^* c_k \langle \psi_j | \psi_k \rangle \right|^{1/2}} = \frac{1}{\left| \sum_k \sum_j c_j^* c_k \delta_{jk} \right|^{1/2}} \tag{7.22}$$

where the last identity follows from the orthonormality of eigenvector elements (equation (7.16)). Applying equation (7.17), the summation over k collapses to only one non-zero term where $j = k$, and simplifies the expression to a single summation. For a normalized composite wavefunction, the eigenvector expansion coefficients are thus constrained to the condition:

$$N = \frac{1}{\left| \sum_j |c_j|^2 \right|^{1/2}} \tag{7.23}$$

As an example we will use the first four particle in a box wavefunctions to form an arbitrary particle in a box eigenfunction. Let us suppose for demonstration purposes that we wish this eigenfunction to contain: 5% $|\psi_1\rangle$, 40% $|\psi_2\rangle$, 30% $|\psi_3\rangle$, and

25% $|\psi_4\rangle$. (A general technique for finding the best-fit coefficients c_j will be explored in chapter 10.) This combination forms an un-normalized wavefunction:

$$|\Phi\rangle = N\big(0.05|\psi_1\rangle + 0.40|\psi_2\rangle + 0.30|\psi_3\rangle + 0.25|\psi_4\rangle\big) \tag{7.24}$$

Applying the normalization condition in equation (7.11) or (7.23) we find:

$$N = \frac{1}{\langle\Phi|\Phi\rangle^{1/2}} = \frac{1}{(0.05^2 + 0.40^2 + 0.30^2 + 0.25^2)^{1/2}} = 1.7817 \tag{7.25}$$

The normalized wavefunction is then:

$$|\Phi\rangle = 0.08\,909|\psi_1\rangle + 0.71\,268|\psi_2\rangle + 0.53\,451|\psi_3\rangle + 0.44\,543|\psi_4\rangle \tag{7.26}$$

Note that the coefficients maintain the correct percentage of each eigenfunction in the expanded form, and that they now fulfill the condition:

$$\sum_j |c'_j|^2 = 1 \tag{7.27}$$

PARALLEL INVESTIGATION: Verify that the linear combination wavefunction: $|\Omega\rangle$ that is 10% $|\chi_1\rangle$, 30% $|\chi_2\rangle$, 25% $|\chi_3\rangle$, and 35% $|\chi_4\rangle$ has the normalized form: $|\Omega\rangle = 0.18732\,|\chi_1\rangle + 0.56195\,|\chi_2\rangle + 0.46829\,|\chi_3\rangle + 0.65561\,|\chi_4\rangle$.

7.3 Projection operators

A useful tool for combination wavefunctions is the *projection operator*:

$$\hat{P}_n = |\psi_n\rangle\langle\psi_n| \tag{7.28}$$

A set of n orthonormal eigenvector elements have n orthonormal projection operators, which obey the condition:

$$\hat{P}_i \cdot \hat{P}_j = |\psi_i\rangle\langle\psi_i|\psi_j\rangle\langle\psi_j| = |\psi_i\rangle\langle\psi_j|\delta_{ij} \tag{7.29}$$

Equation (7.29) shows that projection operators are *idempotent*:

$$\hat{P}_j^2 = \hat{P}_j \cdot \hat{P}_j = |\psi_j\rangle\langle\psi_j|\psi_j\rangle\langle\psi_j| = |\psi_j\rangle\langle\psi_j| = \hat{P}_j \tag{7.30}$$

As the name implies, when projection operator $\hat{P}_j$ acts on a function expanded from an orthonormal basis, it returns the weight of $|\psi_j\rangle$ in the expansion. For example, applying $\hat{P}_3$ to the function described in equation (7.26):

$$\begin{aligned}\hat{P}_3\,|\,\Phi\rangle = {}& 0.08909|\psi_3\rangle\langle\psi_3|\psi_1\rangle + 0.71268|\psi_3\rangle\langle\psi_3|\psi_2\rangle \\ & + 0.53451|\psi_3\rangle\langle\psi_3|\psi_3\rangle + 0.44543|\psi_3\rangle\langle\psi_3|\psi_4\rangle = 0.53451|\psi_3\rangle\end{aligned} \tag{7.31}$$

Projection operator techniques can be applied to transform a collection of linearly dependent functions into an orthonormal set. The method, like so many others in quantum mechanics, is borrowed from linear algebra and is known as Gram–Schmidt orthogonalization. Suppose there exists a set of normalized real functions $|1\rangle, |2\rangle, |3\rangle \ldots$ with non-zero overlap. The first two can be made orthonormal in the following fashion:

$$|2'\rangle = N(|2\rangle - \hat{P}_1 |2\rangle) = N(|2\rangle - |1\rangle\langle 1|2\rangle) \tag{7.32}$$

Equation (7.32) is self-explicit: we are altering function $|2\rangle$ by projecting from it any component of function $|1\rangle$. The projection is weighted by integral $\langle 1|2\rangle$, which measures their degree of coincidence through their overlap S_{12} as defined in equation (7.19). From equation (7.32) it is apparent if $S_{12} = 0$, then $|2\rangle$ is already orthogonal to $|1\rangle$ and $|2'\rangle = |2\rangle$. The parameter N normalizes the new function, as defined in equation (7.11). It is a little easier to follow if the probability density is formed from the left-hand side of equation (7.32) and integrated to unit probability. Using the fact that the functions are also real, we have:

$$1 = N^2(\langle 2|2\rangle - 2 |\langle 1|2\rangle|^2 + |\langle 1|2\rangle|^2) \tag{7.33}$$

Since the original functions were normalized, N is thus:

$$N = \frac{1}{(1 - \langle 1|2\rangle^2)^{1/2}} = \frac{1}{\left(1 - S_{12}^2\right)^{1/2}} \tag{7.34}$$

Notice the difference between equation (7.34) and equation (7.23) caused by negative signs in the expansion. The functions $|1\rangle$ and $|2'\rangle$ are not only both normalized, they are now orthonormal:

$$\langle 1|2'\rangle = N(\langle 1|2\rangle - \langle 1|1\rangle\langle 1|2\rangle) = 0 \tag{7.35}$$

The same process is performed, projecting $|1\rangle$ from the rest of the set $|3\rangle, |4\rangle, |5\rangle \ldots$ The function $|2'\rangle$ is next projected out of $|3'\rangle, |4'\rangle, |5'\rangle \ldots$, for instance;

$$|3''\rangle = N(|3'\rangle - \hat{P}_{2'} |3'\rangle) = N(|3\rangle - |1\rangle\langle 1|3\rangle - |2'\rangle\langle 2'|3'\rangle) \tag{7.36}$$

Note $|3''\rangle$ remains orthogonal to original function $|1\rangle$:

$$\langle 1|3''\rangle = N(\langle 1|3\rangle - \langle 1|1\rangle\langle 1|3\rangle - \langle 1|2'\rangle\langle 2'|3\rangle) = 0 \tag{7.37}$$

and is similarly orthogonal to $|2'\rangle$. The Gram–Schmidt process is continued for the remaining $i > j$ elements from the set, until all are orthonormal. It is easy to envision this in practice carried out by a computer algorithm.

PARALLEL INVESTIGATION: Verify that the normalization constant for $|3''\rangle$ has the form: $N = \frac{1}{(1 - s_{2'3'}^2)^{1/2}}$.

7.4 Expectation values

Eigenfunctions to a particular Hamiltonian fulfilling Born's criteria are postulated to contain all information pertaining to the system they represent, hence can be used to measure any characteristic physical property. An *expectation value* is a statistically extracted quantum mechanical measurement on the wavefunction which returns the most-likely outcome (what you 'expect' to get) when evaluating a particular property. A recipe for expectation values is concocted from the continuous analogue to equation (7.3). It is represented in Dirac notation as:

$$\langle o \rangle = \frac{\langle \psi \mid \hat{O} \mid \psi \rangle}{\langle \psi | \psi \rangle} = \frac{\int \psi^* \hat{O} \psi \, d\tau}{\int \psi^* \psi \, d\tau} \tag{7.38}$$

The above applies to any wavefunction, but those that are normalized have simplified expressions with denominators: $\langle \psi | \psi \rangle = 1$. The bra and ket in the numerator of equation (7.38) are connected by operator: $\hat{O}$. A synopsis of important quantum mechanical operators was provided in chapter 6, table 6.1.

It is required that observables are represented by real expectation values ($\langle o \rangle = \langle o \rangle^*$). Because a wavefunction may be a superposition of states such as equation (7.26), eigenvalues must also be determined from linear operators which obey the requirement:

$$\hat{O}|\Phi\rangle = \sum_j \hat{O} c_j |\psi_j\rangle = \sum_i c_j \hat{O} |\psi_j\rangle \tag{7.39}$$

Finally, operators which return real eigenvalues must produce equivalent results when acting on either the bra or ket function. In the general case, operation on the bra side requires the *adjoint* operator $\hat{O}^\dagger$, and satisfies the criterion:

$$\langle \psi_i | \hat{O} | \psi_j \rangle = \int \psi_i^* (\hat{O} \psi_j) d\tau = o_{ij} \langle \psi_i | \psi_j \rangle = \langle \psi_j | \hat{O}^\dagger | \psi_i \rangle^*$$
$$= \left(\int (\hat{O}^\dagger \psi_i^*) \psi_j \, d\tau \right)^* = o_{ji}^* \langle \psi_j | \psi_i \rangle^* \tag{7.40}$$

If operators representing measurable properties generate real results, complex conjugation in equation (7.40) would cause no change in either eigenvalue. Equation (7.8) implies the residual overlaps are equal as well, meaning the operator must exhibit self-adjoint, or *Hermitian*, behavior: $\hat{O} = \hat{O}^\dagger$. A related consequence of Hermitian operators is that their eigenfunctions form an orthonormal set, equation (7.16): $\langle \psi_i | \psi_j \rangle = \delta_{ij}$.

To draw a connection between the continuous averaging in equation (7.38) and the discrete averaging of equation (7.3), suppose that ψ is an eigenfunction of $\hat{O}$ returning eigenvalue o after operation. The eigenfunction ψ is also returned, which then forms probability density $|\psi|^2$ when multiplied by the bra function. Eigenvalue o corresponds to x_i in equation (7.3), with $P(x)_i$ being the probability density in that volume element. Infinite summation of all such products results in a weighted average property over all space. Take the case of the energy expectation value for our normalized particle in a 1-dimensional box wavefunctions. Over the limits of the box dimensions energy expectation values have the general form:

$$
\langle \psi_n | \hat{T} | \psi_n \rangle = -\frac{\hbar^2}{2m}\frac{2}{L}\int_0^L \sin\left(\frac{n\pi x}{L}\right)\frac{d^2}{dx}\sin\left(\frac{n\pi x}{L}\right)dx
$$
$$
= +\frac{\hbar^2 n^2 \pi^2}{mL^3}\int_0^L \sin^2\left(\frac{n\pi x}{L}\right)dx = \frac{\hbar^2 n^2 \pi^2}{mL^3}\cdot\frac{L}{2} = \frac{\hbar^2 n^2}{8mL^2}
$$

(7.41)

Particle in a box wavefunctions have statistically-averaged energy expectation values of the exact form of the energy eigenvalue found in chapter 6, equation (6.17). After the operator extracts its eigenfunction, the unaltered probability density infinitely-sums to unity because of normalization.

PARALLEL INVESTIGATION: Verify the expectation values: $\langle \chi_n | \hat{T} | \chi_n \rangle = \frac{h^2 n^2}{32mL^2}$ for wavefunctions of the form: $|\chi_n\rangle = \frac{1}{L^{1/2}}\cos\left(\frac{n\pi x}{2L}\right)$ $n = 1, 3, 5, \ldots$ over the range: $-L \leqslant x \leqslant L$. It should now make sense why this wavefunction has ¼ the energy of our original particle in a box wavefunction. The box length in the present case is double the length of the original, and energy depends on the inverse square of the box dimension.

Suppose equation (7.21) is used to construct a function from an orthonormal basis of eigenvectors $|\psi_j\rangle$ which are eigenfunctions to operator $\hat{O}$ with eigenvalues o. The eigenfunction which is formed has expectation value:

$$
\langle \Phi | \hat{O} | \Phi \rangle = \sum_k \sum_j c_j^* c_k \langle \psi_j | \hat{O} | \psi_k \rangle = \sum_k \sum_j c_j^* c_k o_k \langle \psi_j | \psi_k \rangle = \sum_i \sum_j c_j^* c_k o_k \delta_{jk}
$$

(7.42)

As before the Kronecker delta collapses the summation over j resulting in:

$$
\langle \Phi | \hat{O} | \Phi \rangle = \sum_j |c_j|^2 o_j
$$

(7.43)

When the eigenvector elements are eigenfunctions of a particular operator, expectation values computed for any wavefunction constructed from this set is a weighted average of eigenvalues from elements of the basis. The similarity of equation (7.43) to the discrete probability expression (equation (7.3)) should be noted, particularly when the normality condition (equation (7.27)) is compared to equation (7.4). As an example, consider the energy of our normalized expansion wavefunction which we formed from the first four particle in a box solutions (equation (7.26)). Using equations (7.43) and (7.41), the expectation value is:

$$\langle E \rangle_i = \langle \Phi | \hat{T} | \Phi \rangle = \sum_{j=1}^{4} c_{ij}^2 \frac{j^2 h^2}{8mL^2} = 7.7854 \frac{h^2}{8mL^2} \qquad (7.44)$$

PARALLEL INVESTIGATION: Verify the expectation values: $\langle \Omega | \hat{T} | \Omega \rangle = 29.421 \frac{h^2}{32mL^2}$ for the expansion wavefunction of the form: $|\Omega\rangle = 0.18732|\chi_1\rangle + 0.56195|\chi_2\rangle + 0.46829|\chi_3\rangle + 0.65561|\chi_4\rangle$ where the functions $|\chi_n\rangle$ have quantum numbers: $n = 1, 3, 5,$ ... over the range: $-L \leqslant x \leqslant L$.

Suppose instead that wavefunction ψ is a solution to a particular Hamiltonian of the Schrödinger equation but is not an eigenfunction to some other operator, instead upon its action returning a function altered in some way. An expectation value of the property described by this operator can still be found. It results from an infinite summation of the overlap created by the original bra and altered ket wavefunctions. As an example, the normalized 1-dimensional particle in a box eigenvectors in equation (7.13) are used to evaluate the position expectation value. First note that no member of this eigenvector is an eigenfunction of the position operator: $\hat{x} = xx$. Each returns an altered function of x rather than the original eigenfunction, yet the position expectation value of a particle in a box can be calculated with the help of integrals from appendix B:

$$\langle x \rangle = \frac{2}{L} \int_0^L x\sin^2\left(\frac{n\pi x}{L}\right)dx = \frac{L}{2} \qquad (7.45)$$

Equation (7.45) shows the average particle in a box position is independent of quantum number n, or is the same for all eigenvector elements. Taking into account the statistical nature of expectation values, it is not so surprising that the particle on average is located in the middle of the box.

PARALLEL INVESTIGATION: Verify the expectation values: $\langle \chi_n | \hat{x} | \chi_n \rangle = 0$ for the wavefunctions: $|\chi_n\rangle = \frac{1}{L^{1/2}}\cos\left(\frac{n\pi x}{2L}\right)$ with quantum numbers: $n = 1, 3, 5, ...$ over the range: $-L \leqslant x \leqslant L$. This makes physical sense given the box range. It also makes mathematical sense because the integral being evaluated is an overall *odd function* evaluated over symmetric limits. See section 4 of appendix B for a discussion of integral symmetry.

Since the likelihood of being in box increment dx is given by $|\psi_n|^2$ it is instructive to view plots of this quantity, which is done in figure 7.2 for $|\psi_1\rangle$ and $|\psi_4\rangle$ over the

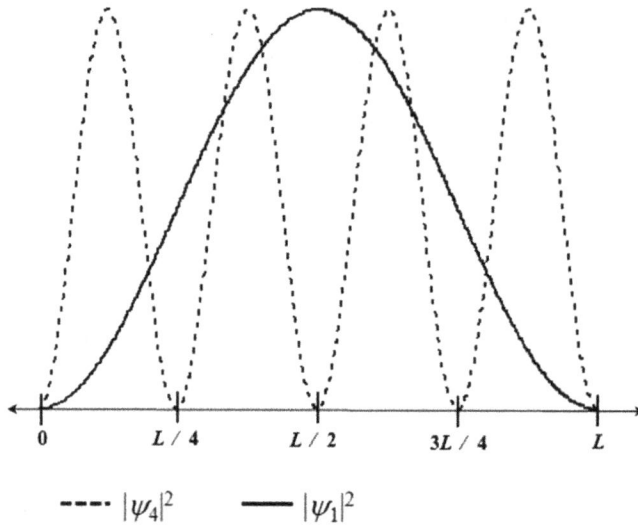

$$---- \; |\psi_4|^2 \qquad \underline{\quad\quad} \; |\psi_1|^2$$

Figure 7.2. Plot of probability density: $|\psi_n|^2$ for eigenvectors: $|\psi_1\rangle = \left(\dfrac{2}{L}\right)^{1/2} \sin\left(\dfrac{\pi x}{L}\right)$ and: $|\psi_4\rangle = \left(\dfrac{2}{L}\right)^{1/2} \sin\left(\dfrac{4\pi x}{L}\right)$ over the range: $0 \leqslant x \leqslant L$.

range: $0 \leqslant x \leqslant L$. These show that indeed $|\psi_1\rangle$ spends most of its time around the center, in fact having maximum probability density exactly at that point. Eigenstate $|\psi_4\rangle$ on the other hand has four evenly distributed regions of probability. Despite that being its statistical average location, the node at $x = L/2$ means a particle described by $|\psi_4\rangle$ in actuality is never at that exact point!

Probability density can be infinitesimally summed over any fraction of its space to find the chance of a particle occupying that region. For instance, the integral: $\int_0^{L/4} |\psi_n|^2 \mathrm{d}x$ gives the prospect of finding a particle in the leftmost quarter of the box. It should be verified using integrals in appendix B or with numeric software that this probability is 0.091 or 9.1% for $|\psi_1\rangle$ and 0.25 or 25% for $|\psi_4\rangle$. Visual estimation based on the plots in figure 7.2 corroborates these results.

Using a set of wavefunctions in an eigenvector it is possible to construct a 2-dimensional array of *matrix elements* to operator $\hat{O}$.

$$o_{ij} = \left\langle \psi_i \,|\hat{O}|\, \psi_j \right\rangle \tag{7.46}$$

Based on the arguments concerning equation (7.40), elements of this matrix must obey the Hermitian property: $o_{ij} = o_{ji}{}^*$. If all basis set elements are eigenfunctions to this operator, the matrix that is formed will be diagonal. Off-diagonal elements are possible if the operator alters the bra or ket function so that some net overlap of functions now occurs over all space.

For example, consider matrix elements of the momentum operator $\hat{p}_x = \dfrac{\hbar}{i}\dfrac{\partial}{\partial x}$ constructed with the 1-dimensional particle in a box eigenvector of equation (7.13) over the range: $0 \leqslant x \leqslant L$. First consider an element along the matrix diagonal.

Applying the momentum operator to eigenvector ψ_4, it should be verified that the integral to be evaluated is:

$$p_{44} = \langle \psi_4 | \hat{p}_x | \psi_4 \rangle = \frac{8\pi\hbar}{iL^2} \int_0^L \sin\left(\frac{4\pi x}{L}\right) \cdot \cos\left(\frac{4\pi x}{L}\right) dx \qquad (7.47)$$

Since the prefactor is imaginary, the Hermitian property is satisfied only if the integral in equation (7.47) is zero. This is verified using the table in appendix B. It is also instructive to visually confirmed this by inspecting the product function: $\sin\left(\frac{4\pi x}{L}\right) \cdot \cos\left(\frac{4\pi x}{L}\right)$ over the range: $0 \leqslant x \leqslant L$, which is displayed in figure 7.3. Again, an axis is provided through the nodal points to aid visualization. The antisymmetric nature of the curve verifies zero overlap for this trigonometric product.

Off-diagonal linear momentum matrix elements will now be both numerically and graphically explored. Using the appropriate wavefunctions and operator, it is straightforward to substantiate the results:

$$p_{34} = \langle \psi_3 | \hat{p}_x | \psi_4 \rangle = \frac{8\pi\hbar}{iL^2} \int_0^L \sin\left(\frac{3\pi x}{L}\right) \cdot \cos\left(\frac{4\pi x}{L}\right) dx = -6.8571\frac{\hbar}{iL} \qquad (7.48)$$

and

$$p_{43} = \langle \psi_4 | \hat{p}_x | \psi_3 \rangle = \frac{6\pi\hbar}{iL^2} \int_0^{2\pi} \sin\left(\frac{4\pi x}{L}\right) \cdot \cos\left(\frac{3\pi x}{L}\right) dx = +6.8571\frac{\hbar}{iL} \qquad (7.49)$$

As required, the matrix elements show the property: $p_{34} = p_{43}{}^*$. For visual demonstration, amplitude plots of the above sine-cosine products are provided in figures 7.4(a) and (b). Although they are not exact mirror images, the area sum under out of phase regions leads to their product amplitudes having opposing signs.

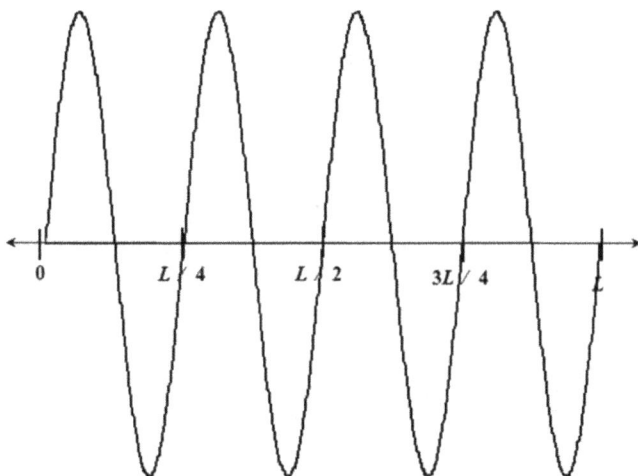

Figure 7.3. Plot of the amplitude product: $\sin\left(\frac{4x}{2}\right) \cdot \cos\left(\frac{4x}{2}\right)$ over the range: $0 \leqslant x \leqslant L$.

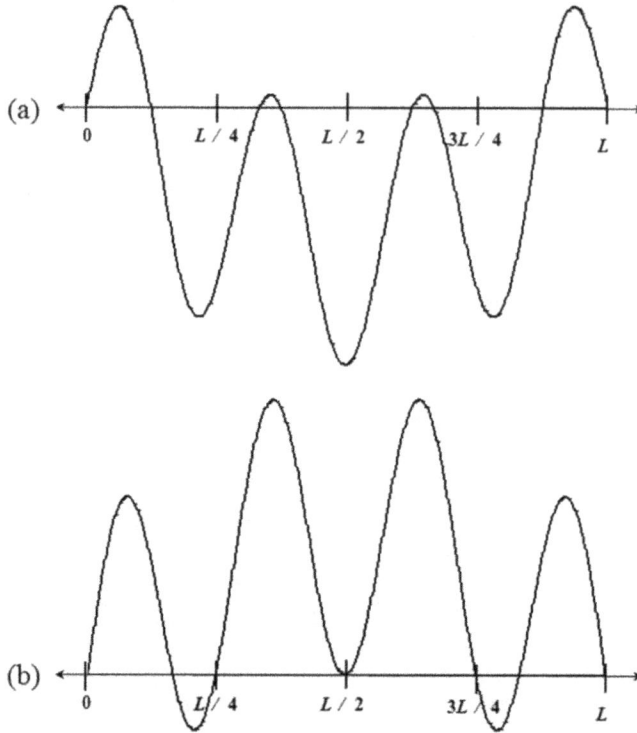

Figure 7.4. (a). Plot of $\frac{8\pi}{L^2}\sin\left(\frac{3\pi x}{L}\right) \cdot \cos\left(\frac{4\pi x}{L}\right)$ vs. x over the range: $.0 \leqslant x \leqslant L$. (b). Plot of $\frac{6\pi}{L^2}\sin\left(\frac{4\pi x}{L}\right) \cdot \cos\left(\frac{3\pi x}{L}\right)$ vs. x over the range: $.0 \leqslant x \leqslant L$.

PARALLEL INVESTIGATION: Verify the matrix elements: $\langle \chi_i | \hat{x} | \chi_j \rangle = 0$ and $\langle \chi_i | \hat{p}_x | \chi_j \rangle = 0$ for the wavefunctions: $|\chi_n\rangle = \frac{1}{L^{1/2}}\cos\left(\frac{n\pi x}{2L}\right)$ with quantum numbers: $n = 1, 3, 5, \ldots$ over the range: $-L \leqslant x \leqslant L$. This makes physical sense given the box range, and mathematical sense, because the integral the first integral is an even × odd × even over symmetric limits, and the second is an even × odd function over symmetric limits (see section B.4).

To further illustrate the utility of matrix elements, let us evaluate the expectation value of position for our expansion wavefunction of equation (7.26). This is accomplished as a sum of individual expectation values weighted by the expansion coefficients:

$$\langle x \rangle = \langle \Phi | \hat{x} | \Phi \rangle = \sum_{j=1}^{4}\sum_{k=1}^{4} c_j c_k \langle \psi_j | \hat{x} | \psi_k \rangle \tag{7.50}$$

The right-hand side of equation (7.42) and equation (7.50) have the same form, but the particle in a box wavefunctions are not eigenfunctions of operator: $\hat{x}$. We therefore cannot extract an eigenvalue, then take advantage of orthogonality and use the Kronecker delta to collapse one of the summations as was done to arrive at equation (7.43). We are forced to explicitly perform a double summation involving several terms to evaluate this property. It is easy to envision this quickly becoming a tiresome process as the size of the expansion set grows.

Despite these complications, the problem can be formulated in a concise, if not elegant, fashion using techniques of linear algebra. Equation (7.50) is symbolically represented in matrix form:

$$\langle x \rangle = \mathbf{C}^{\mathbf{T}} \mathbf{X} \mathbf{C} \tag{7.51}$$

In equation (7.51), $\mathbf{C}$ is a 4×1 column vector with the normalized expansion coefficients from equation (7.26) as elements. The matrix $\mathbf{C}^{\mathbf{T}}$ is the *transpose* of $\mathbf{C}$, which in general means the indices of all row and column elements are interchanged, so the rows become columns and vice versa. It is in fact possible to construct n independent orthonormal expansion vectors from a set of n particle in a box basis eigenvectors, so in general there is possibly an $n \times n$ matrix of coefficients. In this particular case however, $\mathbf{C}^{\mathbf{T}}$ is a 1×4 row vector. If the matrix contained complex values, the elements would not only be transposed, but would necessarily be complex conjugated as well. The combination of transposing and complex conjugating matrix elements is frequently represented with a superscripted dagger:

$$(\mathbf{C}^{\dagger})_{ij} = (\mathbf{C})_{ji}{}^{*} \tag{7.52}$$

Using this notation, Hermitian matrices satisfy the condition: $\mathbf{M}^{\dagger} = \mathbf{M}$. In equation (7.51), $\mathbf{X}$ is a 4×4 matrix with elements: $(\mathbf{X})_{ij} = \langle \psi_i | \hat{x} | \psi_j \rangle$.

Matrix algebra has some commonalities with regular algebra. Matrices may be added or subtracted, provided their dimensionalities are the same. A matrix may also be multiplied or divided by a numeric or *scalar* quantity, as long as the same operation is performed on all elements. It should be noted that multiplication of matrices may be, but is not necessarily, commutative, or in other words it is possible that: $\mathbf{Q} \mathbf{R} \neq \mathbf{R} \mathbf{Q}$. Because of this it is very important to distinguish 'left-multiplying' from 'right-multiplying'. A square matrix with non-zero elements along the diagonal: $(\mathbf{M})_{ii}$, and all others: $(\mathbf{M})_{ij} = 0$, for $i \neq j$ is called a diagonal matrix. The unit matrix: $\mathbf{1}$ is the special case of a diagonal matrix with all 1's along its diagonal. In general, all diagonal matrices commute with any matrix it can multiply. The rules of matrix multiplication are very specific. For this operation the matrices must be *conformable*, meaning their inner indices have the same dimension. In other words the column length of the left matrix must match the row length of the right matrix. The product then has dimensions matching the outer values of the matrices under combination, or the number of rows of the left matrix and columns of the right. For example, an element of the product: $\mathbf{P} = \mathbf{Q} \mathbf{R}$ is found using the recipe:

$$P_{ij} = \sum_{k} Q_{ik} R_{kj} \tag{7.53}$$

The multiplication procedure is thus the summation of products of the left-hand matrix row elements multiplied by the right-hand matrix column elements. For those

$$(1\cdot4)+(-2\cdot2)+(4\cdot1)+(-1\cdot5)$$

$$
\begin{bmatrix}
1 & -2 & 4 & -1 \\
0 & -2 & 2 & 3 \\
1 & -6 & 0 & 2 \\
7 & 2 & 1 & 0
\end{bmatrix}
\times
\begin{bmatrix}
4 & 7 & 3 & -4 \\
2 & -3 & 5 & 1 \\
1 & 0 & 4 & -7 \\
5 & 0 & 3 & -2
\end{bmatrix}
=
\begin{bmatrix}
-1 & 13 & 6 & -32 \\
13 & 6 & 7 & -22 \\
2 & 25 & -21 & -14 \\
33 & 43 & 35 & -33
\end{bmatrix}
$$

Figure 7.5. Performing a matrix multiplication.

unfamiliar with linear algebra, figure 7.5 demonstrates this process for the row 1 column 1 element of a product matrix. The other elements of the product matrix can be computed for your own practice. Matrix division is more appropriately described as multiplication by an inverse. Multiplication of a matrix by its inverse is a commutative process: $\mathbf{M}^{-1}\mathbf{M} = \mathbf{M}\mathbf{M}^{-1} = \mathbf{1}$.

We now are in 'position' to use equation (7.45) to evaluate the average location of a particle described by our expansion wavefunction in a box of length L. The first task is to evaluate elements: $\langle \psi_i | \hat{x} | \psi_j \rangle$ over the first four particle in a box wavefunctions, then to perform the appropriate matrix multiplications with the expansion coefficients. Because the particle in a box solutions are real functions, $\mathbf{X}$ is symmetric, or $X_{ij} = X_{ji}$. In fact, since all numbers are real, complex conjugation has no affect on the elements. Hence this matrix is Hermitian: $\mathbf{X}^\dagger = \mathbf{X}$. Note also that the conformability of the triple matrix product in equation (7.51) is: $(1 \times 4) \cdot (4 \times 4) \cdot (4 \times 1)$ which indeed produces a 1×1, or single-valued result.

Equation (7.51) can be executed in one of two ways. Although matrix algebra is not necessarily commutative it does always obey the associative property: $(\mathbf{Q}\,(\mathbf{R}\,\mathbf{S})) = ((\mathbf{Q}\,\mathbf{R})\,\mathbf{S})$. For no particular reason, we choose to first left multiply the property matrix by the transpose of the coefficients, or evaluate: $\mathbf{Z} = \mathbf{C}^{\mathrm{T}}\mathbf{X}$. The industrious reader can verify the matrices needed and their product is:

$$
\mathbf{Z} = [00.089 \quad 0.713 \quad 0.535 \quad 0.445]
\begin{bmatrix}
0.500L & -0.181L & 0.0 & -0.014L \\
-0.181L & 0.500L & -0.195L & 0.0 \\
0.0 & -0.195L & 0.500L & -0.199 \\
-0.014L & 0.0 & -0.199 & 0.500L
\end{bmatrix}
\tag{7.54}
$$

$$
= [-0.090L \quad 0.236L \quad 0.040L \quad 0.115L]
$$

The property is found by then using $\mathbf{Z}$ in a right multiplication by the coefficient matrix, or $\langle x \rangle_i = \mathbf{Z}\,\mathbf{C}$:

$$
\langle x \rangle = [-0.090L \quad 0.236L \quad 0.040L \quad 0.115L]
\begin{bmatrix}
0.089 \\
0.713 \\
0.535 \\
0.445
\end{bmatrix}
= 0.233L
\tag{7.55}
$$

If instead the calculation was begun by combining the two rightmost matrices: $\mathbf{Z} = \mathbf{X}\,\mathbf{C}$, the result would be a 4×1 column matrix which is the transpose of $\mathbf{Z}$ found in equation (7.55). Left multiplication of this with $\mathbf{C}^{\mathrm{T}}$ would produce the same answer.

The result of equation (7.54) and (7.55) shows that a particle described by equation (7.26) is found on average a little less than one fourth of the way into the box. As visual verification, figure 7.6 provides an amplitude plot of our normalized wavefunction expanded from the first four particle in a box solutions. As additional verification, the function's probability density can be summed for the left-hand quarter of the box:

$$\int_0^{L/4} \Phi^* \Phi \, dx \tag{7.56}$$

Again, the system of equations to solve is: $P = \mathbf{C}^T \mathbf{P} \mathbf{C}$ when cast in matrix form, where $\mathbf{P}$ is the probability density matrix constructed from the first four particle in a box wavefunctions over that range:

$$(P)_{nm} = \frac{2}{L} \int_0^{L/4} \sin\left(\frac{n\pi x}{L}\right) \sin\left(\frac{m\pi x}{L}\right) dx \tag{7.57}$$

With this definition the matrix problem takes the form shown below. The actual matrix product should be performed in two steps as was done in equations (7.54) and (7.55). But for brevity we only display the result after both are completed:

$$P = [0.089 \quad 0.713 \quad 0.535 \quad 0.445] \begin{bmatrix} 0.0908 & 0.1501 & 0.1592 & 0.1200 \\ 0.1501 & 0.2500 & 0.2701 & 0.2122 \\ 0.1592 & 0.2701 & 0.3031 & 0.2572 \\ 0.1200 & 0.2122 & 0.2572 & 0.2500 \end{bmatrix} \begin{bmatrix} 0.089 \\ 0.713 \\ 0.535 \\ 0.445 \end{bmatrix} \tag{7.58}$$

$$= 0.7706$$

Notice that the 11 and 44 elements of the $\mathbf{P}$ matrix are values that were found in our previous discussion of partial summation of $|\psi|^2$. Equation (7.58) shows, according to its probability density sum in that region, a particle described by equation (7.26) has a 77.06% chance of being in the leftmost quarter of the box. Again figure 7.6 corroborates this assertion.

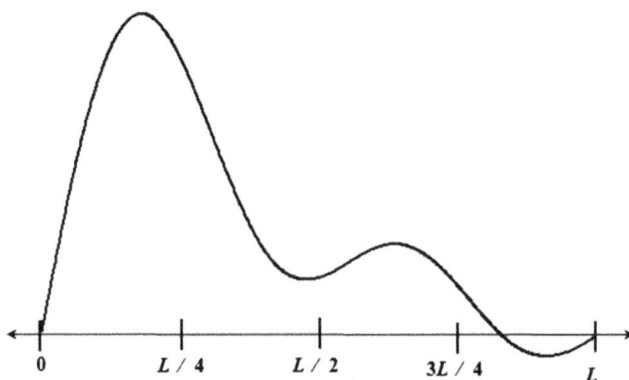

Figure 7.6. Plot of the expansion function: $|\Phi\rangle = 0.08909|\psi_1\rangle + 0.71268|\psi_2\rangle + 0.53451|\psi_3\rangle + 0.44543|\psi_4\rangle$ with $|\psi_1\rangle - |\psi_4\rangle$ taken from the first four normalized particle in a box wavefunctions.

In a similar fashion, the average momentum of Φ_i can be determined by solving the system of equations: $\langle p \rangle_i = \mathbf{C}^T \mathbf{p} \mathbf{C}$. Equations (7.42) and (7.49) show the derivation of matrix elements p_{34} and p_{43}, respectively. In general the numeric values of $\mathbf{p}$ form a 4×4 anti-symmetric matrix, and when the imaginary factor from the momentum operator is included the matrix exhibits the Hermitian property: $\mathbf{p}^\dagger = \mathbf{p}$. The interested reader should complete the evaluation of momentum matrix elements and perform the triple matrix product to find an average momentum for $|\Phi\rangle$ of:

$$\langle p \rangle = [0.089 \quad 0.713 \quad 0.535 \quad 0.445] \begin{bmatrix} 0 & -2.67\dfrac{\hbar}{iL} & 0 & -1.07\dfrac{\hbar}{iL} \\ +2.67\dfrac{\hbar}{iL} & 0 & -4.80\dfrac{\hbar}{iL} & 0 \\ 0 & +4.80\dfrac{\hbar}{iL} & 0 & -6.86\dfrac{\hbar}{iL} \\ +1.07\dfrac{\hbar}{iL} & 0 & +6.86\dfrac{\hbar}{iL} & 0 \end{bmatrix} \begin{bmatrix} 0.089 \\ 0.713 \\ 0.535 \\ 0.445 \end{bmatrix} \quad (7.59)$$

$$= 0$$

Despite the somewhat tedious effort that went into solving this problem, it is nonetheless satisfying that this matrix product is zero. Recall that the functions forming the basis of this expansion are not eigenfunctions of momentum and in fact each have zero average momentum in the box. There is also a philosophical requirement matching this mathematical result. Expectation values are required to match real measurements. If this matrix product had produced an imaginary result, it would be in conflict with the postulates of quantum mechanics.

A final point regarding the matrix formulation for calculating quantum mechanical expectation values. Equations (7.54) and (7.55) represent a generalized recipe for properties of wavefunctions expanded from a linear combination of basis functions. It should be recognized that equation (7.43) represents a specialized case of the same problem where the property matrix constructed from basis eigenfunctions is *diagonal*. In this case orthogonality forces the off-diagonal elements to all be zero. Equation (2.37) only applies when two criteria are met: (1) the basis set elements are eigenfunctions of the operator representing the property being calculated, and (2) the basis set elements form an orthonormal set.

7.5 The uncertainty principle

As matter moves through space, the classical description allows simultaneous knowledge of its position, momentum, kinetic and potential energies at all times during its trajectory, limited only by the accuracy of devices used to measure properties. In quantum mechanics however, it turns out that certain physical parameters have limitations on the precision to which they can simultaneously be known. This limit may be beyond the capability of devices used to make their measurement, however it inherently exists. In this instance, we say there is an *uncertainty* inherent to their simultaneous measurement.

Consider the action of two different quantum mechanical operators consecutively acting on a wavefunction. If $|\psi\rangle$ is a simultaneous eigenfunction of both $\hat{O}_1$ and $\hat{O}_2$, the order of operation is irrelevant, because eigenfunctions o_1 and o_2 are constants so:

$$\hat{O}_1\hat{O}_2\,|\psi\rangle = \hat{O}_1 o_2|\psi\rangle = o_2\hat{O}_1|\psi\rangle = o_2 o_1|\psi\rangle = \hat{O}_2\hat{O}_1|\psi\rangle$$
$$= \hat{O}_2 o_1|\psi\rangle = o_1\hat{O}_2|\psi\rangle = o_1 o_2|\psi\rangle \qquad (7.60)$$

In mathematical terms, operators successively applied to a simultaneous eigenfunction of both exhibit the commutative property. In quantum mechanics, we define the *commutator*, with the definition:

$$[\hat{O}_1,\ \hat{O}_2]\,|\psi\rangle = (\hat{O}_1\hat{O}_2 - \hat{O}_2\hat{O}_1)|\,\psi\rangle \qquad (7.61)$$

If $|\psi\rangle$ is an eigenfunction of both operators:

$$[\hat{O}_1,\ \hat{O}_2]\,|\psi\rangle = 0 \qquad (7.62)$$

However, first applying an operator which alters the wavefunction in some way can possibly, but not always, affect the action of the second. In this case the order of operation can lead to different results when properties are evaluated.

Examples of commuting operators include those for position: $\hat{x}$, $\hat{y}$, and $\hat{z}$. Likewise, the three individual components of linear momentum commute among each other. However, consider the action of any two of these along the same coordinate direction, for instance:

$$[\hat{x},\ \hat{p}_x]|\psi\rangle = \frac{\hbar}{i}\left(x\cdot\frac{\partial}{\partial x}\,|\psi\rangle - \frac{\partial}{\partial x}x\cdot|\psi\rangle\right)$$
$$= \frac{\hbar}{i}\left(x\cdot\left|\frac{\partial\psi}{\partial x}\right\rangle - |\psi\rangle - x\cdot\left|\frac{\partial\psi}{\partial x}\right\rangle\right) = -\frac{\hbar}{i}\,|\psi\rangle \qquad (7.63)$$

The x values of position and linear momentum, and other pairs which do not commute, are known as *conjugate variables*. With regard to angular momentum, note that no individual component commutes with another, for instance using operators from table 6.1:

$$\left[\hat{L}_z,\ \hat{L}_y\right]|\psi\rangle = -\hbar^2\left(\left(x\cdot\frac{\partial}{\partial y} - y\cdot\frac{\partial}{\partial x}\right)\cdot\left(z\cdot\frac{\partial}{\partial x} - x\cdot\frac{\partial}{\partial z}\right)|\,\psi\rangle\right.$$
$$\left.- \left(z\cdot\frac{\partial}{\partial x} - x\cdot\frac{\partial}{\partial z}\right)\cdot\left(x\cdot\frac{\partial}{\partial y} - y\cdot\frac{\partial}{\partial x}\right)|\psi\rangle\right)$$
$$= -\hbar^2\left(\left(\left(xz\cdot\left|\frac{\partial^2\psi}{\partial y\partial x}\right\rangle - x^2\left|\frac{\partial^2\psi}{\partial y\partial z}\right\rangle - yz\left|\frac{\partial^2\psi}{\partial x^2}\right\rangle + y\left|\frac{\partial\psi}{\partial z}\right\rangle + yx\left|\frac{\partial^2\psi}{\partial x\partial z}\right\rangle\right)\right)\right.$$
$$\left.-\left(z\left|\frac{\partial\psi}{\partial y}\right\rangle + zx\cdot\left|\frac{\partial^2\psi}{\partial x\partial y}\right\rangle - zy\left|\frac{\partial^2\psi}{\partial x^2}\right\rangle - x^2\cdot\left|\frac{\partial^2\psi}{\partial z\partial y}\right\rangle + xy\left|\frac{\partial^2\psi}{\partial z\partial x}\right\rangle\right)\right) \qquad (7.64)$$
$$= -\hbar^2\left(y\left|\frac{\partial\psi}{\partial z}\right\rangle - z\left|\frac{\partial\psi}{\partial y}\right\rangle\right) = -i\hbar\hat{L}_x\,|\psi\rangle$$

In 1927 Heisenberg showed that two conjugate variables p and q, which are eigenvalues of operators: $\hat{P}$ and $\hat{Q}$ respectively, have an inherent limit to the precision of their expectation values given by:

$$\Delta p \Delta q \geqslant \frac{\hbar}{2} \tag{7.65}$$

We explore some ramifications of Heisenberg's result in this section.

Earlier in chapter 6 we discussed the free particle, a system experiencing no potential, with wavefunction described by equation (6.9). Translation in the positive direction gives a solution: $\psi = Ae^{+ikx}$. As shown in chapter 6, these functions are eigenfunctions of momentum with eigenvalues: $p_x = \hbar k$, which is an exact result with no variance in its measurement. The wavefunction has probability density:

$$|\psi(x)|^2 = |A|^2 (e^{-ikx}e^{+ikx}) = |A|^2 e^0 = |A|^2 \tag{7.66}$$

which is independent of x, or constant over all space. This means a free particle is equally likely be anywhere, exhibiting no localization whatsoever. There is no statistical precision in knowledge of position. According to equation (7.65), as the degree of uncertainty in linear momentum decreases: $\Delta p_x \rightarrow 0$, the uncertainty in position becomes infinite: $\Delta x \rightarrow \infty$.

In probability theory, statistical variance is a measure of the spread of values from the mean. In order to properly account for both positive and negative deviation, the uncertainty in precision of a statistically-averaged value is found from its variance as a root mean square:

$$\Delta q = \sqrt{\langle q^2 \rangle - \langle q \rangle^2} \tag{7.67}$$

We can use these quantities to investigate consequences of the uncertainty principle for conjugate variables such as position and momentum.

Consider uncertainty in the lowest energy one dimensional particle in a box wavefunction: ψ_1 given in equation (7.13). For these eigenfunctions, general forms for expectation values position are given in equation (7.45) and momentum in equation (7.47). The results were independent of quantum number n, with values: $\langle p_x \rangle = 0$ and $\langle x \rangle = L/2$. What remains to be determined are the following expectation values unique to ψ_1:

$$\langle x^2 \rangle = \frac{2}{L} \int_0^L x^2 \sin^2\left(\frac{\pi x}{L}\right) dx = \frac{2}{L}\left(\frac{L^3}{6} - \frac{L^3}{4\pi^2}\right) = \frac{2\pi^2 L^2 - 3L^2}{6\pi^2} \tag{7.68}$$

and

$$\langle p_x^2 \rangle = -\hbar^2 \frac{2}{L} \int_0^L \sin\left(\frac{\pi x}{L}\right)\frac{d^2}{dx^2}\sin\left(\frac{\pi x}{L}\right)dx = +\frac{2\hbar^2\pi^2}{L^3}\int_0^L \sin^2\left(\frac{\pi x}{L}\right) = \frac{\hbar^2\pi^2}{L^2} \tag{7.69}$$

The position-momentum uncertainty in the lowest energy state of a 1-dimensional particle in a box is then:

$$\Delta x \cdot \Delta p_x = \sqrt{\frac{2\pi^2 L^2 - 3L^2}{6\pi^2} - \frac{L^2}{4}} \cdot \sqrt{\frac{\hbar^2 \pi^2}{L^2} - 0} = \sqrt{\frac{\pi^2 - 6}{3}} \cdot \frac{\hbar}{2}$$

$$= 1.1357\frac{\hbar}{2} \tag{7.70}$$

or 13.6% above the minimum uncertainty. It is left as an exercise to show that the uncertainty in particle in a box eigenstate ψ_2 is: $1.6703\hbar/2$.

PARALLEL INVESTIGATION: Verify the uncertainty between position and momentum is: $\Delta x \cdot \Delta p_x = 1.1357\frac{\hbar}{2}$ for the wavefunction: $|\chi_1\rangle = \frac{1}{L^{1/2}} \cos\left(\frac{\pi x}{2L}\right)$ with quantum numbers over the range: $-L \leqslant x \leqslant L$.

IOP Concise Physics

What's the Matter with Waves?
An introduction to techniques and applications of quantum mechanics
William Parkinson

Chapter 8

Quantum rotation

8.1 Circular motion: the particle on a ring

To review the classical physics of matter in circular motion, see chapter 4. For its quantum mechanical representation we take the same approach as there, developing it from the perspective of angular momentum. For circular motion, the vector describing angular momentum is perpendicular to the plane of rotation, written relative to the radial vector and linear momentum of the particle in the usual way as: $\vec{L} = \vec{r} \times \vec{p}$. If rotation is confined to the xy-plane the z-component of L is the only non-zero value, and has quantum mechanical form that is a combination of position and linear momentum operators:

$$\widehat{L_z} = \hat{x} \cdot \hat{p}_y - \hat{y} \cdot \hat{p}_x = \frac{\hbar}{i}\left[x \cdot \frac{\partial}{\partial y} - y \cdot \frac{\partial}{\partial x}\right] \tag{8.1}$$

As was found to be the case in chapter 4, the development is facilitated by transforming coordinate systems from Cartesian into the spherical polar frame. This is accomplished using the relations (see figure 4.2)

$$z = r \cos\theta \qquad x = r \sin\theta \cos\phi \qquad y = r \sin\theta \sin\phi$$

$$r = (x^2 + y^2 + z^2)^{1/2} \quad \theta = \cos^{-1}\left(\frac{z}{r}\right) \qquad \phi = \tan^{-1}\left(\frac{y}{x}\right) \tag{8.2}$$

Expressions from section B.1 for derivatives of inverse trigonometric functions will also be needed:

$$\frac{d\cos^{-1}(y)}{dx} = -\frac{1}{\sqrt{1 - y^2}} \cdot \frac{dy}{dx} \qquad \frac{d\tan^{-1}(y)}{dx} = \frac{1}{1 + y^2} \cdot \frac{dy}{dx} \tag{8.3}$$

doi:10.1088/978-1-6817-4577-0ch8
8-1

Coordinate conversion is accomplished by performing transformations of the following type:

$$\frac{\partial}{\partial x} = \left(\frac{\partial r}{\partial x}\right)_{y,z} \frac{\partial}{\partial r} + \left(\frac{\partial \theta}{\partial x}\right)_{y,z} \frac{\partial}{\partial \theta} + \left(\frac{\partial \phi}{\partial x}\right)_{y,z} \frac{\partial}{\partial \phi} \tag{8.4}$$

with similar equations for $\partial/\partial y$ and $\partial/\partial z$. Beginning with the radial coordinate r as defined in equation (8.2) we find:

$$\left(\frac{\partial r}{\partial x}\right)_{y,z} = \frac{1}{2}(x^2 + y^2 + z^2)^{-1/2} \cdot 2x = \frac{x}{r} = \sin \theta \cos \phi \tag{8.5}$$

The final identity follows by substituting the spherical polar definition of x from equation (8.2) into equation (8.5). Similarly the y dependence of r is:

$$\left(\frac{\partial r}{\partial y}\right)_{x,z} = \frac{y}{r} = \sin \theta \sin \phi \tag{8.6}$$

Proceeding for θ we have:

$$\left(\frac{\partial \theta}{\partial x}\right)_{y,z} = -\frac{1}{\sqrt{1 - (z/r)^2}} \cdot -\left(\frac{1}{2}\right)\frac{2zx}{r^3} = \frac{1}{\sqrt{(x^2 + y^2)/r^2}} \cdot \frac{zx}{r^3} \tag{8.7}$$

Substituting definitions from equation (8.2) for x, y, and z, equation (8.7) simplifies to:

$$\left(\frac{\partial \theta}{\partial x}\right)_{y,z} = \frac{\cos \theta \cos \phi}{r} \tag{8.8}$$

In a similar fashion, we obtain:

$$\left(\frac{\partial \theta}{\partial y}\right)_{x,z} = \frac{\cos \theta \sin \phi}{r} \tag{8.9}$$

Proceeding for ϕ we have:

$$\left(\frac{\partial \phi}{\partial x}\right)_{y,z} = \frac{1}{1 + (y/x)^2} \cdot -\frac{y}{x^2} = -\frac{1}{(x^2 + y^2)/x^2} \cdot \frac{y}{x^2} \tag{8.10}$$

Again substituting the spherical polar definitions and simplifying gives:

$$\left(\frac{\partial \phi}{\partial x}\right)_{y,z} = -\frac{\sin \phi}{r \sin \theta} \tag{8.11}$$

The motivated reader can verify that:

$$\left(\frac{\partial \phi}{\partial y}\right)_{x,z} = \frac{\cos \phi}{r \sin \theta} \tag{8.12}$$

These transformations are combined:

$$\frac{\partial}{\partial x} = \sin\theta\cos\phi \cdot \frac{\partial}{\partial r} + \frac{\cos\theta\cos\phi}{r} \cdot \frac{\partial}{\partial\theta} - \frac{\sin\phi}{r\sin\theta} \cdot \frac{\partial}{\partial\phi}$$

$$\frac{\partial}{\partial y} = \sin\theta\sin\phi \cdot \frac{\partial}{\partial r} + \frac{\cos\theta\sin\phi}{r} \cdot \frac{\partial}{\partial\theta} + \frac{\cos\phi}{r\sin\theta} \cdot \frac{\partial}{\partial\phi}$$

(8.13)

so that the z-component of angular momentum is:

$$\hat{L}_z = \frac{\hbar}{i} \cdot r\sin\theta\cos\phi\left[\sin\theta\sin\phi \cdot \frac{\partial}{\partial r} + \frac{\cos\theta\sin\phi}{r} \cdot \frac{\partial}{\partial\theta} + \frac{\cos\phi}{r\sin\theta} \cdot \frac{\partial}{\partial\phi}\right]$$

$$- \frac{\hbar}{i} \cdot r\sin\theta\sin\phi\left[\sin\theta\cos\phi \cdot \frac{\partial}{\partial r} + \frac{\cos\theta\cos\phi}{r} \cdot \frac{\partial}{\partial\theta} - \frac{\sin\phi}{r\sin\theta} \cdot \frac{\partial}{\partial\phi}\right]$$

(8.14)

In similar fashion to the classical investigation in chapter 4, this cumbersome quantum mechanical expression simplifies nicely to a z-component of angular momentum in concise form under the restrictions of constant radial component and constant polar angle of 1.571 radians:

$$\hat{L}_z = \frac{\hbar}{i}\frac{\partial}{\partial\phi}$$

(8.15)

As discussed in chapter 4, the energy of rotation is: $E = L^2/2I$, with moment of inertia: $I = \mu r^2$. For rotational motion confined to the xy-plane: $L^2 = L_z^2$ only. The time independent Schrödinger equation for a particle rotating in the xy-plane under the influence of no external potential is therefore a function of the azimuthal angle:

$$\frac{\hat{L}_z^2}{2I}\psi(\phi) = -\frac{\hbar^2}{2I}\frac{\partial^2\psi(\phi)}{\partial\phi^2} = E\psi(\phi)$$

(8.16)

This standard quantum mechanical problem is commonly referred to as the *particle on a ring*. We seek eigenfunctions whose second derivative returns the eigenvalue: $-2IE/\hbar^2$. This problem should be familiar by now, with solutions of either sine, cosine or imaginary exponential form. In this particular case, it is typical to employ wavefunctions that are also eigenfunctions of angular momentum. The preference is:

$$\psi(\phi) = Ne^{im_\ell\phi}$$

(8.17)

Prefactor N normalizes the function, determined by requiring the probability density infinitely sums to unity over the range: $0 \leqslant \phi \leqslant 2\pi$. Because the wavefunction is imaginary, N is determined from the integral:

$$1 = \int_0^{2\pi} |\psi(\phi)|^2 \, d\phi = N^2\int_0^{2\pi} e^{-im_\ell\phi}e^{+im_\ell\phi} \, d\phi = N^2\int_0^{2\pi} d\phi = N^2 \cdot 2\pi$$

(8.18)

The normalized wavefunction is therefore: $\psi(\phi) = 1/\sqrt{2\pi} \cdot e^{im_\ell\phi}$

The parameter m_ℓ is of more immediate interest. Using equations (8.16) and (8.17), we see that the energy of the particle on a ring has m_ℓ dependence:

$$E_{m_\ell} = \frac{m_\ell^2 \hbar^2}{2I} \qquad (8.19)$$

This same result could equally be obtained via the energy expectation value: $\langle E \rangle = \langle \psi(\phi)|\hat{T}|\psi(\phi)\rangle$ over the range: $0 \leqslant \phi \leqslant 2\pi$. From equation (8.15) we see that angular momentum is also a function of this parameter:

$$\hat{L}_z\psi(\phi) = \frac{\hbar}{i}\frac{1}{\sqrt{2\pi}}\frac{\partial e^{im_\ell\phi}}{\partial\phi} = m_\ell\hbar\psi(\phi) \qquad (8.20)$$

Alternatively, the result from (8.20) can be found by evaluating: $\langle L_z \rangle = \langle \psi(\phi)|\hat{L}_z|\psi(\phi)\rangle$. To understand why an eigenfunction with negative exponential was not considered in the general solution, we allow the parameter m_ℓ to take either positive or negative values, giving angular momentum oriented along either the positive or negative z axis. The two possibilities arise when a particle rotates in opposite directions. Since m_ℓ appears as a squared factor in equation (8.19), both directions of motion have the same energy, or are degenerate.

At this point there are no restrictions on the numeric values that m_ℓ can possess, so that both angular momentum and rotational energy appear continuous. However a suitable wavefunction must meet the stipulation that as it traverses the ring it does not destructively interfere with itself. In order to fulfill this condition, its amplitude must periodically align. In terms of angular displacement, the boundary condition to be satisfied is: $\psi(\pm m_\ell \cdot 2\pi) = \psi(0)$, for integer values of m_ℓ. To visualize this restriction, think of the real component of the exponential represented in the form: $\cos(m_\ell\phi) + i\sin(m_\ell\phi)$. Focusing on the real component, it is then required that $(\cos(0) = \cos \pm m_\ell \times 2\pi)$. As an alternative we can think from Bohr's perspective of a well-behaved wavefunction. They are postulated to be single-valued, so the particle on a ring must be at the same point in space every 2π radians. From either perspective, the boundary condition is satisfied if m_ℓ is restricted to values:

$$m_\ell = 0, \pm 1, \pm 2, \pm 3... \qquad (8.21)$$

The parameter m_ℓ is thus a quantum number. Angular momentum and energy exist in quantized, or discrete, amounts. There are both positive and negative allowed values of angular momentum, scaled by integer units of $\hbar$. The sign of m_ℓ reflects opposing directions of particle rotation. Energy levels for the particle on a ring are depicted in figure 8.1. Those above the zero-point are doubly-degenerate since m_ℓ appears as a square in the energy expression.

The particle on a ring wavefunction contains much the same type of information that we saw could be extracted from the particle in a box expression in chapter 7. For instance, the likelihood that the particle be found in the third quadrant is determined by evaluating:

$$\frac{1}{2\pi}\int_\pi^{3\pi/2} e^{-im_\ell\phi}e^{+im_\ell\phi}\,\mathrm{d}\phi = \frac{1}{2\pi}\int_\pi^{3\pi/2}\mathrm{d}\phi = \frac{1}{2\pi}\cdot\phi\,\bigg|_\pi^{3\pi/2} = \frac{3\pi/2 - \pi}{2\pi} = 0.25 \quad (8.22)$$

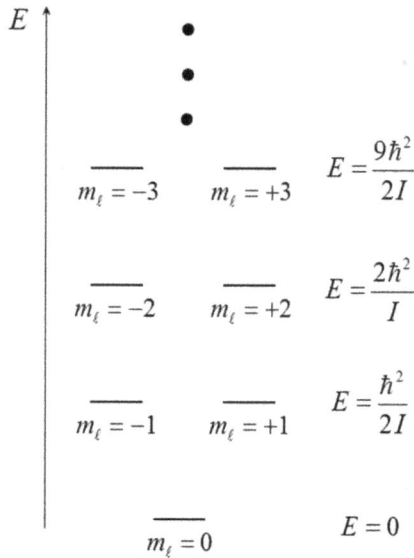

Figure 8.1. Energy levels of a particle on a ring.

Indeed the particle on a ring has 25% of being in any quadrant. In fact it has the same probability of being in any arbitrarily chosen ring segment independent of m_ℓ value. In chapter 7 we saw that the particle in a box had a likelihood of being in a box segment which depended upon its quantum number n. An expectation value for the most probable angular position can be determined by designing the operator: $\hat{\phi} = \phi\times$. Applying this to the normalized particle on a ring wavefunction we find:

$$\langle\phi\rangle = \langle\psi(\phi)|\hat{\phi}|\psi(\phi)\rangle = \frac{1}{2\pi}\int_0^{2\pi} \phi \ \mathrm{d}\phi = \frac{1}{4\pi}\phi^2 \ \bigg|\begin{matrix}2\pi\\0\end{matrix} = \pi \qquad (8.23)$$

Referring back to chapter 7, this is analogous to the particle in a box. The particle on a ring on average is halfway around, independent of quantum number.

8.2 Spherical motion: the particle on a sphere

A particle confined to rotational motion on the surface of a sphere will now experience a time independent Schrödinger equation in Cartesian form:

$$\frac{\hat{L}^2}{2I}\psi(x, y, z) = E\psi(x, y, z) \qquad (8.24)$$

where: $\hat{L}^2 = \hat{L}_x^2 + \hat{L}_y^2 + \hat{L}_z^2$. Again, it will be advantageous to choose the spherical polar coordinate frame. Since motion is now explicitly 3-dimensional, additional components of angular momentum are required. Similar transformations to those for $\hat{L}_z$ result in spherical polar expressions for the other two angular momentum components:

$$\hat{L}_x = -\frac{\hbar}{i}\left[\sin\phi\frac{\partial}{\partial\theta} + \cos\phi\frac{\cos\theta}{\sin\theta}\frac{\partial}{\partial\phi}\right]$$

$$\hat{L}_y = \frac{\hbar}{i}\left[\cos\phi\frac{\partial}{\partial\theta} - \sin\phi\frac{\cos\theta}{\sin\theta}\frac{\partial}{\partial\phi}\right]$$

(8.25)

The three independent components are combined via vector dot product to form the total angular momentum: $\hat{L}^2 = \hat{L}_x^2 + \hat{L}_y^2 + \hat{L}_z^2$. After some manipulation this quantity is expressed as:

$$\hat{L}^2 = -\hbar^2\left[\frac{\partial^2}{\partial\theta^2} + \frac{\cos\theta}{\sin\theta}\frac{\partial}{\partial\theta} + \frac{1}{\sin^2\theta}\frac{\partial^2}{\partial\phi^2}\right]$$

(8.26)

The $\hat{L}^2$ operator is used in equation (8.24), solving the eigenvalue problem for eigenfunctions of both the polar and azimuthal angle: $\psi(\theta, \phi)$. Following a tedious separation of variables which is omitted here but can be found in a variety of mathematical physics texts, eigenfunctions of the $\hat{L}^2$ operator are identified. Collectively the solutions are known as the *spherical harmonics*, introduced by LaPlace in the 18th century. These versatile functions are applicable to a wide variety of problems that possess spherical symmetry. Their utility spreads across the disciplines of physics; not only to quantum mechanics but classical mechanics and electrostatics as well.

As we learned from the 2- and 3-dimensional particle in a box in chapter 6, separation of variables produces a product function of one term for each independent variable. In spherical harmonics, dependence on the azimuthal angle is expressed in the familiar particle on a ring solutions: $e^{im_\ell\phi}$, containing the quantum number m_ℓ. Polar angle dependence is characterized by trigonometric functions known as *associated Legendre polynomials*. This requires an additional quantum number symbolized as: ℓ, which recursively generates the spherical harmonic's Legendre polynomials. Acceptable solutions result when ℓ takes the integer values:

$$\ell = 0, 1, 2, \ldots$$

(8.27)

The existence of ℓ creates a boundary condition which places upper limits on allowed values for m_ℓ:

$$m_\ell = 0, \pm 1, \pm 2, \ldots, \pm\ell$$

(8.28)

Spherical harmonics are symbolically represented by: $Y_\ell^{m_\ell}(\theta, \phi)$. Forms of spherical harmonics for $\ell = 0$ to $\ell = 3$ are presented in table 8.1. These functions form an orthonormal set over the polar range: $0 \leqslant \theta \leqslant \pi$ and the azimuthal range: $0 \leqslant \phi \leqslant 2\pi$. When evaluating properties over the entirety of their space, care

Table 8.1. Spherical harmonic expressions for $\ell = 0$ to $\ell = 3$.

ℓ	m_ℓ	$Y_\ell^{m_\ell}(\theta, \phi)$
0	0	$\frac{1}{2\sqrt{\pi}}$
1	0	$\frac{1}{2}\sqrt{\frac{3}{\pi}}\cos(\theta)$
1	± 1	$\mp\frac{1}{2}\sqrt{\frac{3}{2\pi}}\sin(\theta)\cdot e^{\pm i\phi}$
2	0	$\frac{1}{4}\sqrt{\frac{5}{\pi}}(3\cos^2(\theta) - 1)$
2	± 1	$\mp\frac{1}{2}\sqrt{\frac{15}{2\pi}}\cos(\theta)\cdot\sin(\theta)\cdot e^{\pm i\phi}$
2	± 2	$\frac{1}{4}\sqrt{\frac{15}{2\pi}}\sin^2(\theta)\cdot e^{\pm 2i\phi}$
3	0	$\frac{1}{4}\sqrt{\frac{7}{\pi}}(2\cos(\theta) - 3\cos(\theta)\cdot\sin(\theta))$
3	± 1	$\mp\frac{1}{8}\sqrt{\frac{21}{\pi}}(4\cos^2(\theta)\cdot\sin(\theta) - \sin^3(\theta))\cdot e^{\pm i\phi}$
3	± 2	$\frac{1}{4}\sqrt{\frac{105}{2\pi}}\cos(\theta)\cdot\sin^2(\theta)\cdot e^{\pm 2i\phi}$
3	± 3	$\mp\frac{1}{8}\sqrt{\frac{35}{\pi}}\sin^3(\theta)\cdot e^{\pm 3i\phi}$

must be taken in correctly expressing the angular integration element, which includes a sine factor for projection of the azimuthal angle into the xy plane:

$$\int d\tau = \int_0^{2\pi} d\phi \cdot \int_0^\pi \sin(\theta)d\theta \tag{8.29}$$

Using equation (8.29), table 8.1, and integrals of section B.2, we see for instance the $Y_2^0(\theta, \phi)$ function is normalized:

$$\langle Y_2^0(\theta, \phi)| Y_2^0(\theta, \phi)\rangle = \frac{5}{16\pi}\int_0^{2\pi} d\phi \cdot \int_0^\pi \sin(\theta)[9\cos^4(\theta) - 6\cos^2(\theta) + 1]d\theta$$

$$= \frac{5}{8}\left[-\frac{9}{5}\cos^5(\theta) + 2\cos^3(\theta) - \cos(\theta)\right] \tag{8.30}$$

$$\Big|_0^{2\pi} = \frac{5}{8}\left[\frac{18 - 20 + 10}{5}\right] = 1$$

and is orthogonal to for instance the $Y_1^0(\theta, \phi)$ function:

$$\langle Y_2^0(\theta, \phi)| Y_1^0(\theta, \phi)\rangle = \frac{15}{8\pi}\int_0^{2\pi} d\phi \cdot \int_0^\pi \sin(\theta)[3\cos^3(\theta) - \cos(\theta)]d\theta$$

$$= \frac{15}{4}\left[-\frac{3}{4}\cos^4(\theta) + \frac{1}{2}\cos^2(\theta)\right] \tag{8.31}$$

$$\Big|_0^{2\pi} = \frac{15}{4}\left[-\frac{(-1)^4 - (1)^4}{4} + \frac{(-1)^2 - (1)^2}{2}\right] = 0$$

Energy eigenvalues for the particle on a sphere wavefunctions have the general form:

$$E_\ell = \frac{\ell(\ell + 1)\hbar^2}{2I} \tag{8.32}$$

For instance, applying equation (8.24) on the $\ell = 2$, $m_\ell = 0$ spherical harmonic function, along with trigonometric identity A2.3.1 leads to:

$$\hat{H}Y_2^0(\theta, \phi) = -\frac{1}{4}\sqrt{\frac{5}{\pi}}\frac{\hbar^2}{2I}\left[\frac{\partial^2}{\partial\theta^2} + \frac{\cos\theta}{\sin\theta}\frac{\partial}{\partial\theta} + \frac{1}{\sin^2\theta}\frac{\partial^2}{\partial\phi^2}\right](3\cos^2(\theta) - 1)$$

$$= -\frac{1}{4}\sqrt{\frac{5}{\pi}}\frac{\hbar^2}{2I}(6\sin^2(\theta) - 12\cos^2(\theta)) \tag{8.33}$$

$$= -\frac{1}{4}\sqrt{\frac{5}{\pi}}\frac{\hbar^2}{2I}(6 - 18\cos^2(\theta)) = \frac{2(2+1)\hbar^2}{2I} \cdot Y_2^0(\theta, \phi)$$

If instead, this value is determined as an expectation value, the eigenfunction would be extracted from the eigenvalue and then using equation (8.32) we would find:

$$\langle Y_2^0(\theta, \phi)|\hat{H}|Y_2^0(\theta, \phi)\rangle = \frac{2(2+1)\hbar^2}{2I} \cdot \langle Y_2^0(\theta, \phi)|Y_2^0(\theta, \phi)\rangle = \frac{2(2+1)\hbar^2}{2I} \tag{8.34}$$

The spherical harmonic wavefunctions represent particles possessing total angular momentum: $\sqrt{\ell(\ell + 1)}\,\hbar$. It is important to notice that both the particle on a sphere energy and total angular momentum depend on ℓ alone and have no m_ℓ dependence. Since equation (8.28) gives the number of allowed m_ℓ values as a function of ℓ, the particle on a sphere energy levels therefore exhibit a $2\ell + 1$ degeneracy depicted in figure 8.2.

As discussed for a particle on a ring, m_ℓ measures the z-projection of angular momentum in units: $L_z = m_\ell\hbar$. Allowed values of m_ℓ (equation (8.28)), determine

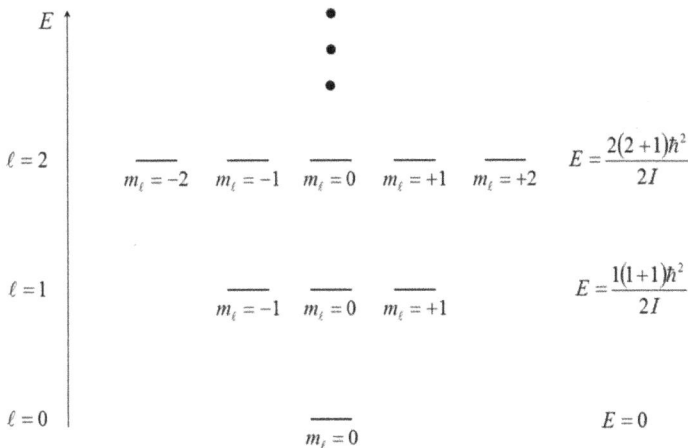

Figure 8.2. Energy levels of a particle on a sphere.

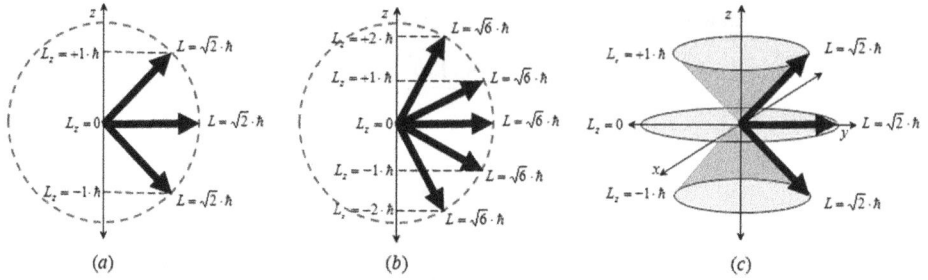

Figure 8.3. (*a*) Vector picture of one-particle angular momentum for a p-type orbital ($\ell = 1$). (*b*) Vector picture of one-particle angular momentum for a d-type orbital ($\ell = 2$). (*c*) Precessing vector picture of one-particle angular momentum for a p-type orbital ($\ell = 1$).

the degeneracy of a given orbital on each energy level. For a given ℓ, the L_z value can take $2\ell + 1$ possible orientations, commonly depicted by the vector picture shown in figure 8.3. Using the approach taken in section 7.5 it is possible to show the commutation relation: $[\hat{L}^2, L_z] = 0$. Spherical harmonics are therefore simultaneous eigenfunctions of both operators. The relationships: $[\hat{L}^2, L_y] = 0$, and $[\hat{L}^2, L_x] = 0$ are also true. But chapter 7, equation (7.64) shows that individual components are of angular momentum are conjugate variables which do not commute with one another, so cannot be simultaneously measured. According to equation (7.65), if $\langle \hat{L}_z \rangle$ is known with no variance, it is impossible to determine $\langle \hat{L}_x \rangle$ or $\langle \hat{L}_y \rangle$ with any certainty at all. This consequence is indicated by the *precessing vector* model of angular momentum shown in figure 8.3(c).

IOP Concise Physics

What's the Matter with Waves?
An introduction to techniques and applications of quantum mechanics
William Parkinson

Chapter 9

Quantum vibration

9.1 Harmonic oscillation

The task of investigating quantum mechanical vibration benefits from a wealth of information in chapter 3, a classical study of this motion. It may be helpful to review that treatment or refer back when needed, as there are terms and concepts common to both models. For the present study we take a familiar approach, introduced in chapter 2 and also used in chapter 3, by treating two vibrating objects as a one-body oscillator characterized by reduced mass μ. The coordinate frame is aligned so the motion occurs 1-dimensionally along the x-axis. As in chapter 3, we assume the mass experiences restoring force: $\vec{F} = -k\vec{x}$. There k was defined as the spring constant with SI units: $kg \cdot s^{-2}$. That term in this context conjures an image of a nano-spring attached to the mass. Although the name spring constant invokes a concept of some merit, here we prefer k to be known as the *force constant*. Its origin is electrostatic not mechanical, more correctly envisioned as a characterizing factor to the potential gradient rather than spring stiffness (for mathematical confirmation of this assertion, see section 9.2 and particularly equation (9.25)). Ultimately the magnitude of k is indeed proportional to the strength of a chemical bond between two nuclei.

Equation (3.5) shows a recipe for determining the above-mentioned connection between a 1-dimensional conservative restoring force and its potential. This is expressed in quantum mechanical form by an operator equation:

$$\hat{V} = \frac{1}{2}k\hat{x}^2 \qquad (9.1)$$

Equation (9.1) adds a layer of complication to the time independent Schrödinger equation. The quantum mechanical investigations in chapters 6 and 8 had no terms beyond a kinetic energy operator. The eigenvalue problem is now a differential equation of the form:

$$\hat{H} \mid \psi \rangle = \left(-\frac{\hbar^2}{2m}\frac{d^2}{dx^2} + \frac{1}{2}kx^2 \right) \mid \psi \rangle = E|\psi\rangle \qquad (9.2)$$

Equation (9.2) shares mathematical similarities to the eigenvalue problem discussed in section 8.2 for a particle on a sphere. In that case, the Hamiltonian contained no explicit potential term, but the kinetic energy operator expressed in spherical polar coordinates resulted in a differential equation with coefficients that are functions of independent variable θ (see equation (8.26)). As was the case there, the differential equation has a power series solution resulting in an eigenfunction formed from a product of an exponential term and a polynomial. This time, however, the exponential is of Gaussian form. In addition, the associated Legendre polynomials used in chapter 8 are replaced by Hermite polynomials, symbolized: $H_v(y)$. Expressions for the first eight Hermite polynomials are presented in table 9.1. Solution of the differential equation of equation (9.2) can be found in a variety of mathematical physics texts. The resulting harmonic oscillator eigenfunctions have a normalized form:

$$|\psi_v\rangle = N_v H_v(y) \cdot e^{-y^2/2} \quad v = 0, 1, 2, \ldots \tag{9.3}$$

The prefactor N_v normalizes the functions:

$$N_v = \frac{1}{(2^v \cdot v!)^{1/2}}\left(\frac{\alpha}{\pi}\right)^{1/4} \tag{9.4}$$

and the parameters y and α are given by:

$$y = \sqrt{\alpha} \cdot x \quad \alpha = \frac{\mu\omega}{\hbar} \tag{9.5}$$

where ω is the angular speed, having SI units of $\mathrm{rad} \cdot \mathrm{s}^{-1}$ and defined in chapter 3, equation (3.9):

$$\omega = \sqrt{\frac{k}{\mu}} \tag{9.6}$$

Table 9.1. The first eight Hermite polynomials.

v	$H_v(y)^\dagger$
0	1
1	$2y$
2	$4y^2 - 2$
3	$8y^3 - 12y$
4	$16y^4 - 48y^2 + 12$
5	$32y^5 - 160y^3 + 120y$
6	$64y^6 - 480y^4 + 720y^2 - 120$
7	$128y^7 - 1344y^5 + 3360y^3 - 1680y$

$\dagger y = \sqrt{\alpha} \cdot x \; \alpha^2 = \mu\omega/\hbar$

The parameter α has SI dimension: m^{-2}. The oscillator frequency is typically represented by the Greek symbol ν instead of f, and is determined by dividing the angular speed by a complete oscillator cycle of 2π radians:

$$\nu = \frac{\omega}{2\pi} = \frac{1}{2\pi}\sqrt{\frac{k}{\mu}} \tag{9.7}$$

This is measured in SI units of s^{-1} or Hertz. The energy eigenvalues are expressed using either the angular speed:

$$E_v = (v + 1/2)\hbar\omega \quad v = 0, 1, 2, \ldots \tag{9.8}$$

or oscillator frequency:

$$E_v = (v + 1/2)h\nu \quad v = 0, 1, 2, \ldots \tag{9.9}$$

As encountered for rotational motion, harmonic oscillator energy levels are quantized by integer amounts, with solutions beginning at level zero. Unlike the rotational case, the $v = 0$ solution has a non-zero value:

$$E_0 = \frac{1}{2}\hbar\omega = \frac{1}{2}h\nu \tag{9.10}$$

Equation (9.10) shows the harmonic oscillator *zero-point vibrational energy*, which it is argued occurs as a consequence of the uncertainty principle (see section 7.5). The implication is that the oscillator must remain in motion even in its definite state of minimum energy, otherwise its position could be precisely determined along with its conjugate momentum. Similar arguments are posed in the science of cryogenics as to why absolute zero cannot experimentally be attained. Another distinctive feature of equation (9.9) is the energy level spacing. All harmonic oscillator energy levels from zero to infinity are separated by the same amount: $\hbar\omega = h\nu$. This fact was used in section 3.2 to determine an expression for the spectral brightness of a black body radiator.

Normalization of $|\psi_0\rangle$ is demonstrated using equations (9.3) and (9.4) along with table 9.1. Integrating its probability density over the limits $-\infty \leqslant x \leqslant +\infty$, we obtain:

$$\langle\psi_0|\psi_0\rangle = \left(\frac{\alpha}{\pi}\right)^{1/2} \int_{-\infty}^{+\infty} e^{-\alpha x^2}\mathrm{d}y \tag{9.11}$$

The integration is performed on an even function over symmetric limits in equation (9.11). According to section B.4, it therefore can be evaluated in the form:

$$\langle\psi_0|\psi_0\rangle = 2\cdot\left(\frac{\alpha}{\pi}\right)^{1/2} \int_0^{+\infty} e^{-\alpha x^2}\mathrm{d}y = 2\cdot\left(\frac{\alpha}{\pi}\right)^{1/2} \times \frac{1}{2}\left(\frac{\pi}{\alpha}\right)^{1/2} = 1 \tag{9.12}$$

PARALLEL INVESTIGATION: Verify that $v = 1$ harmonic wavefunction is normalized: $\langle\psi_0|\psi_0\rangle = 1$ over the range: $-\infty \leqslant x \leqslant +\infty$.

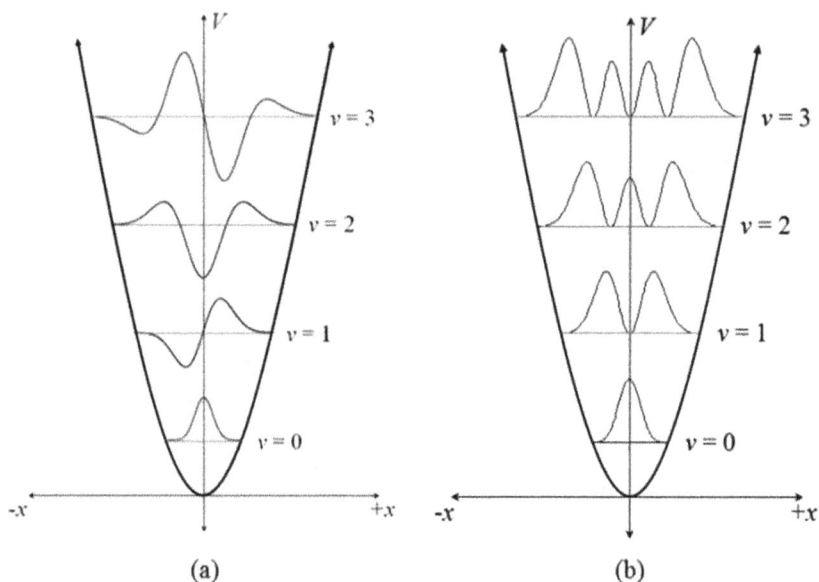

Figure 9.1. (a) Amplitude plots of the $v = 0$, 1, 2, and 3 harmonic oscillator wavefunctions. (b) Probability density plots of the $v = 0$, 1, 2, and 3 harmonic oscillator wavefunctions.

Plots of the first four harmonic oscillator wavefunction amplitudes in figure 9.1(a) are characteristic in appearance for all higher order harmonic oscillator wavefunctions. Although integration extends to infinity, their amplitudes show that the probability density of $|\psi_0\rangle$ to be very localized about the zero-point displacement. For instance using a typical diatomic force constant such as that for Cl_2 of $k = 329$ kg $\cdot$ s^{-2} (= 0.211 au), we find using numerical software the ground state oscillator has a 66.2% chance of being within a Bohr radius, or: $\pm a_0$ of $x = 0$. This increases to 94.5% within $\pm 2a_0$ and 99.6% within $\pm 3a_0$.

> **PARALLEL INVESTIGATION:** Verify that harmonic oscillator wavefunction: $|\psi_1\rangle$ has a 41% chance of being in the region of space: $a_0 \leqslant x \leqslant +\infty$.

According to appendix B, harmonic oscillator wavefunctions with even-ordered Hermite polynomials must be orthogonal to those with odd order over all space $-\infty \leqslant x \leqslant +\infty$. In fact, all others also exhibit this property. For instance:

$$\langle\psi_0|\psi_2\rangle = \frac{1}{2\sqrt{2}}\left(\frac{\alpha}{\pi}\right)^{1/2} \cdot \left[8\alpha\int_0^\infty x^2 e^{-\alpha x^2}dx - 4\int_0^\infty e^{-\alpha x^2}dx\right]$$
$$= \frac{1}{2\sqrt{2}}\left(\frac{\alpha}{\pi}\right)^{1/2} \cdot \left[8\alpha \cdot \frac{\sqrt{\pi}}{4\alpha^{3/2}} - 4\frac{1}{2}\left(\frac{\pi}{\alpha}\right)\right] = 0$$

(9.13)

PARALLEL INVESTIGATION: Verify: $\langle \psi_1 | \psi_3 \rangle = 0$ over the range: $-\infty \leqslant x \leqslant +\infty$.

The wavefunctions are eigenfunctions of the Hamiltonian defined in equation (9.2). For instance, using $| \psi_1 \rangle$:

$$-\frac{\hbar^2}{2\mu}\frac{d^2}{dx^2}[N_1(2\sqrt{\alpha}x)e^{-\alpha x^2/2}] + \frac{1}{2}kx^2N_1(2\sqrt{\alpha}x)e^{-\alpha x^2/2} = E_1N_1(2\sqrt{\alpha}x)e^{-\alpha x^2/2} \quad (9.14)$$

Differentiating and dividing through by $|\psi_1\rangle$ leaves an eigenvalue of:

$$E_1 = \frac{3\alpha\hbar^2}{2\mu} - \frac{\hbar^2\alpha^2}{2\mu}x^2 + \frac{k}{2}x^2 \quad (9.15)$$

Substitution of equations (9.5) and (9.6) into equation (9.15) gives:

$$E_1 = \frac{3}{2}\hbar\omega = (1 + 1/2)\hbar\omega \quad (9.16)$$

PARALLEL INVESTIGATION: Verify the expectation value: $\langle \psi_1 | \hat{H} | \psi_1 \rangle = (1 + 1/2)\hbar\omega$ for the normalized harmonic oscillator wavefunction $|\psi_1\rangle$ over the range: $-\infty \leqslant x \leqslant +\infty$.

As further examples of harmonic oscillator expectation value problems, we now use the ground state wavefunction $| \psi_0 \rangle$ to calculate the uncertainty between linear position and momentum. From equation (9.10), the minimum energy point of the harmonic oscillator is not zero, but has a finite value in accordance with Heisenberg's relationship. From the variance recipe described in section 7.5, the oscillator has uncertainty between position and momentum described by:

$$\Delta x \cdot \Delta p_x = \sqrt{\langle x^2 \rangle - \langle x \rangle^2} \cdot \sqrt{\langle p_x^2 \rangle - \langle p_x \rangle^2} \quad (9.17)$$

The integral required to evaluate average position is:

$$\langle x \rangle = \langle \psi_0 | \hat{x} | \psi_0 \rangle = \left(\frac{\alpha}{\pi}\right)^{1/2} \int_{-\infty}^{+\infty} x \cdot e^{-\alpha x^2}dx = 0 \quad (9.18)$$

The integral in equation (9.18) can be evaluated with numeric software to obtain zero, but this conclusion can easily be drawn by noting that integration is performed on an even $\times$ odd $\times$ even = odd function over symmetric limits. According to section B.4, this integral must vanish. Likewise, the expectation value of linear momentum is evaluated from:

$$\langle p_x \rangle = \langle \psi_0 | \hat{p}_x | \psi_0 \rangle = -\frac{\hbar\alpha}{i}\left(\frac{\alpha}{\pi}\right)^{1/2} \int_{-\infty}^{+\infty} x \cdot e^{-\alpha x^2}dx = 0 \quad (9.19)$$

which vanishes for the same reason as does equation (9.18).

What remain to be determined are the expectation values of even × even × even = even functions, which are done using integrals from appendix B:

$$\langle x^2 \rangle = \langle \psi_0 \mid \hat{x}^2 \mid \psi_0 \rangle = 2 \cdot \left(\frac{\alpha}{\pi}\right)^{1/2} \int_0^{+\infty} x^2 \cdot e^{-\alpha x^2} \, \mathrm{d}x = 2 \cdot \left(\frac{\alpha}{\pi}\right)^{1/2} \frac{\sqrt{\pi}}{4\alpha^{3/2}} = \frac{1}{2\alpha} \quad (9.20)$$

Notice from equation (9.5) that the result of equation (9.20) has dimension m^2, as required. The remaining integral to evaluate is:

$$\langle p_x^2 \rangle = \langle \psi_0 \mid \hat{p}_x^2 \mid \psi_0 \rangle = -2\hbar^2 \cdot \left(\frac{\alpha}{\pi}\right)^{1/2} \left[-\alpha \int_0^{+\infty} e^{-\alpha x^2} \mathrm{d}x + \alpha^2 \int_0^{+\infty} x^2 \cdot e^{-\alpha x^2} \mathrm{d}x \right]$$

$$= -2\hbar^2 \cdot \left(\frac{\alpha}{\pi}\right)^{1/2} \left[-\alpha \frac{1}{2}\sqrt{\frac{\pi}{\alpha}} + \alpha^2 \frac{\sqrt{\pi}}{4\alpha^{3/2}} \right] = \frac{\hbar^2 \alpha}{2} \quad (9.21)$$

Checking the dimension of equation (9.21), we find: kg$^2 \cdot$ m$^2 \cdot$ s^{-2}, as required. Inserting equations (9.20) and (9.21) into equation (9.17) gives:

$$\Delta x \cdot \Delta p_x = \sqrt{1/2\alpha - 0} \cdot \sqrt{\hbar^2 \alpha / 2 - 0} = \frac{\hbar}{2} \quad (9.22)$$

or the minimum uncertainty between conjugate variables for the ground state harmonic oscillator.

PARALLEL INVESTIGATION: Verify the uncertainty between position and momentum of the first excited harmonic oscillator wavefunction: $|\psi_1\rangle$ is: $\Delta x \cdot \Delta p_x = 3\hbar/2$ over the range: $-\infty \leqslant x \leqslant +\infty$.

9.2 Anharmonicity

Applying the ideas of section 9.1 to molecules has severe limitations. The potential of two nuclei vibrating about their equilibrium bond length: r_e, as in a diatomic, is only parabolic at very small displacements. If we consider a parabola with its zero point placed at r_e, we soon encounter problems at points $r < r_e$ due to the increase in steepness of the potential caused by internuclear repulsion. In addition at values $r > r_e$, there eventually is bond breaking. To mimic this event, the potential should approach zero as the nuclei asymptotically approach infinite separation, rather than a potential exhibiting continued parabolic increase. To address these issues, the potential can be expanded in a Taylor series about the point $r - r_e$:

$$V(r) = V(r_e) + \left(\frac{\mathrm{d}V(r)}{\mathrm{d}r}\right)_{r=r_e} (r - r_e) + \frac{1}{2!}\left(\frac{\mathrm{d}^2 V(r)}{\mathrm{d}r^2}\right)_{r=r_e} (r - r_e)^2 + \cdots \quad (9.23)$$

The potential is adjusted with the first term in the expansion set at the origin so that $V(r_e) = 0$. At the bottom of the well $V(r)$ varies with r such that there is zero slope with only upward curvature at the inflection point $r = r_e$. Therefore, through the quadratic term the power series is:

$$V(r) \approx \frac{1}{2!}\left(\frac{d^2 V(r)}{dr^2}\right)_{r=r_e} (r - r_e)^2 \qquad (9.24)$$

Terms beyond second order in the power series expansion are called *anharmonic effects*. Using equation (9.1), we can identify the force constant from equation (9.24):

$$k = \left(\frac{d^2 V(r)}{dr^2}\right)_{r=r_e} \qquad (9.25)$$

Based on probability density plots of harmonic oscillator wavefunctions shown in figure 9.1(b), we see the ground state to be highly localized about the point $r = r_e$. It is thus anticipated the harmonic potential from equation (9.1) or (9.24) is reasonable for the $v = 0$ wavefunction. However, effects due to inter-nuclear repulsion and bond breaking become more important as the energy levels increase. Another approach to incorporating anharmonic effects in the oscillator is to replace the parabolic potential with one that more correctly mimics behavior at displacements away from r_e. The most successful of these is the *Morse potential*:

$$V(r) = D_e(1 - e^{-a(r-r_e)})^2 \qquad (9.26)$$

In equation (9.26) D_e is a parameter describing the well depth, alternatively presented in units of Joules, kcal $\cdot$ mol^{-1}, or cm^{-1} depending on the audience. The parameter: a is:

$$a = \left(\frac{\mu\omega^2}{2D_e}\right)^{1/2} \qquad (9.27)$$

Dimensional analysis shows a has units of: m^{-1} as required. The particular form of a is judiciously chosen to correlate behavior of the Morse potential relative to the standard harmonic oscillator potential. This can be seen if the exponential in equation (9.26) is expanded in a power series, to obtain:

$$V(r) = D_e\left(1 - \left(1 - a(r - r_e) + \frac{1}{2!}(a(r - r_e))^2 - \cdots\right)\right)^2 \qquad (9.28)$$

For small displacements: $r - r_e$ only the first term in the expansion is important. We can use equations (9.27) and (9.6) to give:

$$V(r) \approx D_e\left(\left(\frac{\mu\omega^2}{2D_e}\right)^{1/2}(r - r_e)\right)^2 = \frac{1}{2}k(r - r_e)^2 \qquad (9.29)$$

This shows parameter a to be defined so that the Morse potential is harmonic through first order in nuclear displacement.

It is possible to obtain analytic eigenfunctions to the Schrödinger equation's differential equation with the potential in equation (9.26) replacing a parabolic potential in the Hamiltonian. Again a power series solution that is omitted here results, with eigenfunctions containing an exponential instead of a Gaussian. The solution also requires Laguerre polynomials rather than Hermite polynomials. Key features of the Morse potential are presented in figure 9.2. To assist its interpretation, the Morse potential is overlayed with a parabola, reflecting the appearance of a harmonic potential.

At distances: $r < r_e$ where nuclear repulsion is expected to dominate, $V(r)$ displays an enhanced steepness in comparison to a harmonic potential. When $r > r_e$ the Morse potential also displays the proper characteristics with regard to bond breaking. Another result of physical significance to the solutions is the number of energy levels obtained, and their spacing. First there is a finite rather than infinite number of bound state solutions. In addition, the energy levels are no longer evenly spaced as was found in the harmonic oscillator case. In fact, there is a convergence of the energy levels as the bond dissociation limit is met. The quantity D_0 is the bond dissociation energy measured relative to the zero-point vibrational energy. Eigenvalues of the Schrödinger equation containing the Morse potential have the form:

$$E_v = (v + 1/2)\hbar\omega - (v + 1/2)^2 x_e \hbar\omega \quad v = 0, 1, 2, \ldots \tag{9.30}$$

In equation (9.30) x_e is a dimensionless parameter known as the *anharmonicity constant*:

$$x_e = \frac{a^2\hbar}{2\mu\omega} = \frac{\hbar\omega}{4D_e} \tag{9.31}$$

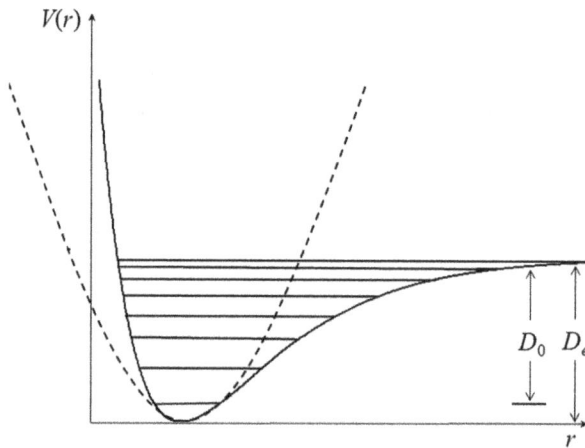

Figure 9.2. The Morse potential: $V(r) = D_e(1 - e^{-a(r-r_0)})^2$ is shown by the solid curve. The converging bound state energy levels are depicted in the potential well. Overlayed with a dashed curve is the harmonic oscillator potential: $V(r) = \frac{1}{2}k(r - r_0)^2$.

Equation (9.30) hints at a truncated power series expansion of a general solution. Experimentalists are thus motivated to express the Morse potential in a generally empirical form including further anharmonic corrections:

$$E_v = (v + 1/2)\hbar\omega - (v + 1/2)^2 x_e\hbar\omega + (v + 1/2)^3 y_e\hbar\omega + \cdots \qquad (9.32)$$

What's the Matter with Waves?

An introduction to techniques and applications of quantum mechanics

William Parkinson

Chapter 10

Variational methods

10.1 Prologue

In 1930, Dirac published his landmark textbook: *Principles of Quantum Mechanics.* With regard to the feasibility of quantum mechanics as a tool for routine application, he stated:

> The underlying physical laws necessary for the mathematical theory of a large part of physics and the whole of chemistry are thus completely known, and the difficulty is only that the exact application of these laws leads to equations much too complicated to be soluble. It therefore becomes desirable that approximate practical methods of applying quantum mechanics should be developed, which can lead to an explanation of the main features of complex atomic systems without too much computation.

The two most common and utile methods employed to address Dirac's concerns are variational and perturbative approaches. Each has distinct advantages and particular applicability to certain quantum mechanical situations. In the following, we examine the basic premise of the variational principle, along with a few simple applications. See chapter 12 for an introduction to perturbative approaches.

10.2 The variational principle

In chapter 6 we learned of the existence of the wavefunction $|\Psi_0\rangle$ that together with Hamiltonian $\hat{H}$ in the time independent Schrödinger equation determine the ground state energy of a particular system of interest:

$$\hat{H}|\Psi_0\rangle = E_0|\Psi_0\rangle \tag{10.1}$$

doi:10.1088/978-1-6817-4577-0ch10

In fact, as discussed in section 7.4, $\hat{H}$ is a Hermitian operator with an eigenvector of orthonormal eigenfunctions $|\Psi_i\rangle$ having a discrete set of energy eigenvalues with the property:

$$E_0 \leqslant E_1 \leqslant E_2 \leqslant \qquad (10.2)$$

The equal sign of equation (10.2) holds in the case of degenerate eigenfunctions. In this instance with no loss of generality we assume the set is non-degenerate, so the equal sign can be ignored.

Suppose an arbitrary function, called the *trial wavefunction*: $|\tilde{\Psi}\rangle$ is formed as a linear combination of the eigenvector:

$$|\tilde{\Psi}\rangle = \sum_i c_i |\Psi_i\rangle \qquad (10.3)$$

Using the projection operator technique (see section 7.3, equation (7.28)), equation (10.3) is written:

$$|\tilde{\Psi}\rangle = \sum_i |\Psi_i\rangle\langle\Psi_i|\tilde{\Psi}\rangle \qquad (10.4)$$

Assuming the trial wavefunction is normalized, the variational principle states that its energy obeys the condition:

$$\langle\tilde{\Psi}|\hat{H}|\tilde{\Psi}\rangle \geqslant E_0 \qquad (10.5)$$

The equality in equation (10.5) holds only in the case: $|\tilde{\Psi}\rangle = |\Psi_0\rangle$. To show this, we begin with the normalization condition on the trial wavefunction, twice insert equation (10.4), and use the orthonormality of the basis:

$$1 = \langle\tilde{\Psi}|\tilde{\Psi}\rangle = \sum_{i,j}\langle\tilde{\Psi}|\Psi_i\rangle\langle\Psi_i|\Psi_j\rangle\langle\Psi_j|\tilde{\Psi}\rangle = \sum_{i,j}\langle\tilde{\Psi}|\Psi_i\rangle\delta_{ij}\langle\Psi_j|\tilde{\Psi}\rangle = \sum_i |\langle\Psi_i|\tilde{\Psi}\rangle|^2 \qquad (10.6)$$

As intimidating as the right-hand side of equation (10.6) may appear, it is merely the sum of the squares of expansion coefficients, in accordance with the normality condition of any function formed from an orthonormal basis (see chapter 7, equation (7.27)). Referring there again, the energy of the trial wavefunction can also be expressed by projection techniques:

$$\langle\tilde{\Psi}|\hat{H}|\tilde{\Psi}\rangle = \sum_{i,j}\langle\tilde{\Psi}|\Psi_i\rangle\langle\Psi_i|\hat{H}|\Psi_j\rangle\langle\Psi_j|\tilde{\Psi}\rangle = \sum_i E_i |\langle\Psi_i|\tilde{\Psi}\rangle|^2 \qquad (10.7)$$

Assuming the basis is not completely degenerate, then each: $E_i > E_0$ according to equation (10.2). This means:

$$\langle\tilde{\Psi}|\hat{H}|\tilde{\Psi}\rangle \geqslant \sum_i E_0 |\langle\Psi_i|\tilde{\Psi}\rangle|^2 \geqslant E_0 \sum_i |\langle\Psi_i|\tilde{\Psi}\rangle|^2 \geqslant E_0 \qquad (10.8)$$

where the final identity follows from equation (10.6). The equal sign in equation (10.8) holds only in the case that $|\tilde{\Psi}\rangle = |\Psi_0\rangle$.

As a general rule, the variational principle is applicable whether or not elements of the expansion eigenvector are eigenfunctions to the Hamiltonian. As an example, consider the function $|\tilde{\psi}\rangle$ as an approximation to the particle in a box wavefunction in the form:

$$|\tilde{\psi}\rangle = \left(\frac{30}{L^5}\right)^{1/2}(xL - x^2) \tag{10.9}$$

This is obviously not an eigenfunction of the kinetic energy operator: $\hat{T}$. The prefactor in equation (10.9) ensures the function is normalized over the range: $0 \leqslant x \leqslant L$:

$$\langle\tilde{\psi}|\tilde{\psi}\rangle = \frac{30}{L^5}\int_0^L (x^2L^2 - 2x^3L + x^4)dx = \frac{30}{L^5}\left(\frac{x^3}{3}L^2 - \frac{x^4}{2}L + \frac{L^5}{5}\right)\bigg|_0^L$$
$$= \frac{30}{L^5}\left(\frac{L^5}{30}\right) = 1 \tag{10.10}$$

PARALLEL INVESTIGATION: Verify the factor: $N = \left(\frac{15}{16L^5}\right)^{1/2}$ normalizes the trial wavefunction: $|\tilde{\chi}\rangle = N \cdot (L^2 - x^2)$ over the range: $-L \leqslant x \leqslant L$.

To illustrate the variational principle, we evaluate the energy expectation value of the wavefunction in equation (10.9):

$$\langle\tilde{E}\rangle = \langle\tilde{\psi}|\hat{T}|\tilde{\psi}\rangle = -\frac{\hbar^2}{2m}\frac{30}{L^5}\int_0^L (xL - x^2)\frac{d}{dx}(xL - x^2)dx$$
$$= \frac{30\hbar^2}{mL^5}\int_0^L (xL - x^2)dx \tag{10.11}$$

Integration and simplification gives:

$$\langle\tilde{E}\rangle = \frac{30\hbar^2}{mL^5}\left(\frac{L^3}{6}\right) = \frac{10}{\pi^2}\frac{h^2}{8mL^2} = 1.0132\frac{h^2}{8mL^2} \tag{10.12}$$

This value is 1.32% above the energy eigenvalue E_1 for particle in a box solution $|\psi_1\rangle$ (see chapter 7, equation (7.41)). Figure 10.1 compares amplitudes for both $|\psi_1\rangle$ and $|\tilde{\psi}\rangle$ over the range: $0 \leqslant x \leqslant L$. We see this node-less function would not be an appropriate upper bound to any of the higher particle in a box solutions.

PARALLEL INVESTIGATION: Verify the energy for the normalized trial wavefunction: $|\tilde{\chi}\rangle = \left(\frac{15}{16L^5}\right)^{1/2}(L^2 - x^2)$ is: $\langle\tilde{E}\rangle = 1.0132\frac{h^2}{32mL^2}$ or 1.32% above the exact solution for the lowest energy state: $|\chi_1\rangle$ over the range: $-L \leqslant x \leqslant L$.

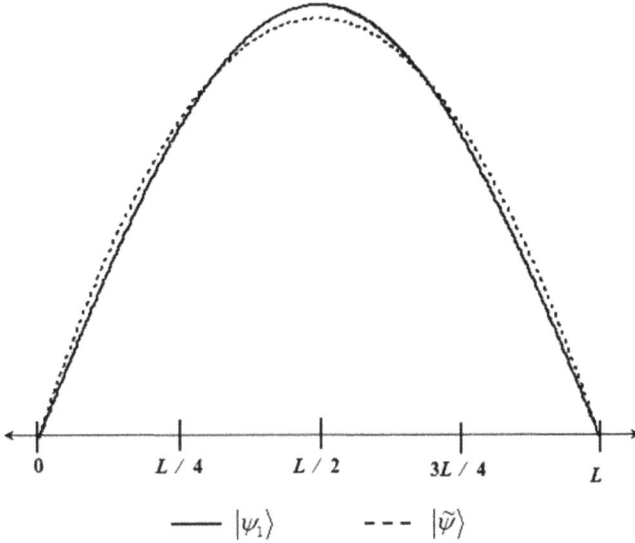

Figure 10.1. Comparing amplitudes of the exact particle in a box wavefunction: $|\psi_1\rangle = \left(\frac{2}{L}\right)^{1/2} \sin\left(\frac{\pi x}{L}\right)$ to the approximate wavefunction: $|\tilde{\psi}\rangle = \left(\frac{30}{L^5}\right)^{1/2}(xL - x^2)$.

As a further example of its utility, consider the variational treatment of a Gaussian function in the form:

$$|\tilde{\psi}\rangle = Ne^{-cx^2} \tag{10.13}$$

as a trial wavefunction for solution to the ground state harmonic oscillator. This problem was discussed in chapter 9, and possesses wavefunctions which are products of a Gaussian function with a Hermite polynomial (see table 9.1). The oscillator energy has the form:

$$\langle E_v \rangle_{\text{exact}} = (v + 1/2)\hbar\omega = (v + 1/2)\hbar\sqrt{\frac{k}{\mu}} \quad v = 0, 1, 2, \ldots \tag{10.14}$$

where k is the oscillator force constant and μ the reduced mass. First consider the normalization of the trial wavefunction over the range: $-\infty \leqslant x \leqslant +\infty$. Since a Gaussian is an even function (see section B.4), we find using integrals from section B.3:

$$\langle \tilde{\psi}|\tilde{\psi}\rangle = 1 = 2N^2 \int_0^\infty e^{-2cx^2}dx = 2N^2 \cdot \frac{1}{2}\sqrt{\frac{\pi}{2c}} \tag{10.15}$$

from which we obtain: $N = (2c/\pi)^{1/4}$. Using the harmonic oscillator Hamiltonian from chapter 9, equation (9.2), the trial energy is:

$$\langle \tilde{E} \rangle = \langle \tilde{\psi}|\hat{H}|\tilde{\psi}\rangle = 2 \cdot \left(\frac{2c}{\pi}\right)^{1/2} \int_0^\infty e^{-cx^2}\left(-\frac{\hbar^2}{2\mu}\frac{d^2}{dx^2} + \frac{1}{2}kx^2\right)e^{-cx^2}dx$$

$$= 2 \cdot \left(\frac{2c}{\pi}\right)^{1/2}\left[\frac{\hbar^2}{2\mu}\sqrt{\frac{\pi c}{2}} - \frac{\hbar^2}{4\mu}\sqrt{\frac{\pi c}{2}} + \frac{k}{16}\sqrt{\frac{\pi c^3}{2}}\right] = \frac{\hbar^2 c}{2\mu} + \frac{k}{8c} \tag{10.16}$$

Our task is to now find the best-fit parameter c. According to the variational principle, the energy expectation value of trial wavefunction $|\tilde{\psi}\rangle$ is greater than or equal to, but not lower than, that for the ground state harmonic oscillator. In mathematical terms, $\langle \tilde{E} \rangle$ is an *upper bound* to $\langle E \rangle_{exact}$. Since we are guaranteed to never go below the exact energy, we are free to optimize c to produce the lowest possible $\langle \tilde{E} \rangle$. Considered as a continuous function, a plot of the trial energy versus c is at a minimum when the tangential slope of this curve is zero, or when:

$$\frac{\partial \langle \tilde{E} \rangle}{\partial c} = 0 \tag{10.17}$$

We therefore use the result of equation (10.16) to find the optimum value of c:

$$\frac{\partial \langle \tilde{E} \rangle}{\partial c} = 0 = \frac{\hbar^2}{2\mu} - \frac{k}{8c^2} \tag{10.18}$$

from which: $c = \sqrt{k\mu}/2\hbar$. When this is inserted into the final identity of equation (10.16) we obtain a trial energy of:

$$\langle \tilde{E} \rangle = \frac{\hbar^2 \sqrt{km}}{4\mu\hbar} + \frac{2k\hbar}{8\sqrt{k\mu}} = \frac{1}{2}\hbar\sqrt{\frac{k}{\mu}} \tag{10.19}$$

which is the exact ground state energy E_0 for the harmonic oscillator according to equation (10.14). Comparing with chapter 9, equation (9.5) we can now see that the optimal value we have determined for the variational parameter is: $c = \alpha/2$. It should therefore be expected to obtain the exact ground state energy because we variationally treated the exact wavefunction (see table 9.1). It is interesting to note this result did not require solving a second-order differential equation with variable coefficients, as does the harmonic oscillator eigenvalue problem.

PARALLEL INVESTIGATION: Verify the un-normalized trial wavefunction: $|\tilde{\psi}\rangle = Nxe^{-cx^2}$ used with the harmonic oscillator Hamiltonian results in the same best-fit variational coefficient found from equation (10.18): $c = \sqrt{k\mu}/2\hbar$ and gives a trial energy: $\langle \tilde{E} \rangle = 3/2\sqrt{k/\mu} \cdot \hbar$. This is the exact energy of the first excited state oscillator: E_1 because as table 9.1 shows, we are varying the exact first excited harmonic oscillator wavefunction.

10.3 Determining expansion coefficients

The most common application of the variational principle is in the determination of an optimal set of coefficients used as weighting factors for a basis set expansion of a trial wavefunction:

$$| \tilde{\Phi}\rangle = N \sum_{j=1}^{n} c_j |\psi_j\rangle \tag{10.20}$$

Equation (10.20) contains N as a normalization constant. For simplicity in the bookkeeping, we will assume both the expansion coefficients and eigenvector elements are real. Using the expectation value recipe from chapter 7, equation (7.38), the trial wavefunction has energy:

$$\langle \tilde{E}\rangle = \frac{\langle \tilde{\Phi} | \hat{H} | \tilde{\Phi}\rangle}{\langle \tilde{\Phi}|\tilde{\Phi}\rangle} = \frac{\displaystyle\sum_{i=1}^{n}\sum_{j=1}^{n} c_i c_j \langle \psi_i | \hat{H} | \psi_j\rangle}{\displaystyle\sum_{i=1}^{n}\sum_{j=1}^{n} c_i c_j \langle \psi_i|\psi_j\rangle} \tag{10.21}$$

The desired outcome is the same as for the harmonic oscillator example above, but it is now necessary to find n best-fit coefficients to optimize the entire expansion. We now have a multi-dimension variational problem, which assures that $\langle \tilde{E}\rangle$ is an upper bound to $\langle E\rangle_{\text{exact}}$. We should keep in mind that $\langle \tilde{E}\rangle$ is an incredibly nuanced energy functional. To graphically represent its dependence on the expansion coefficients requires an $n+1$-dimensional plot. If we sat atop its highest vantage point and looked out at the landscape, we might think we were viewing a mountain vista. The energy surface is pock-marked by a wide array of dips and divots and mounds and mountains as the trial energy rises and falls at the whim of the c_i. There are many routes downward to shallow basins and dales, what we would call *local minima*. In actuality, we seek the deepest chasm, or *global minimum*. Let us choose to take one of the many possible trails down, the quantity: $\langle \tilde{E}_{c_k}\rangle$, which is how the trial energy varies relative to the particular coefficient: c_k. According to the plot of figure 10.2, the slice of the trial energy curve that is singularly a function of c_k is at a minimum when the tangential slope of this curve is zero, or when:

$$\frac{\partial \langle \tilde{E}_{c_k}\rangle}{\partial c_k} = 0 \tag{10.22}$$

Our task is thus to partially differentiate equation (10.21) with respect to a single coefficient c_k. This has the effect of collapsing a summation down to only the term containing that coefficient. From the fact that $\partial c_k / \partial c_k = 1$ and that all coefficients and basis functions are real, we can interchange variables to give two equivalent single summation terms from the differentiation of each double sum. Applying this process to both the numerator and denominator of the expression, according to the product rule, gives:

$$\frac{\partial \langle \tilde{E}_{c_k}\rangle}{\partial c_k} = 0 = \frac{2\displaystyle\sum_{i=1}^{n} c_i \langle \psi_i | \hat{H} | \psi_k\rangle}{\displaystyle\sum_{i=1}^{n}\sum_{j=1}^{n} c_i c_j \langle \psi_i|\psi_j\rangle} - \frac{\displaystyle\sum_{i=1}^{n}\sum_{j=1}^{n} c_i c_j \langle \psi_i|\hat{H}|\psi_j\rangle}{\left[\displaystyle\sum_{i=1}^{n}\sum_{j=1}^{n} c_i c_j \langle \psi_i|\psi_j\rangle\right]^2} 2\displaystyle\sum_{i=1}^{n} c_i \langle \psi_i|\psi_k\rangle \tag{10.23}$$

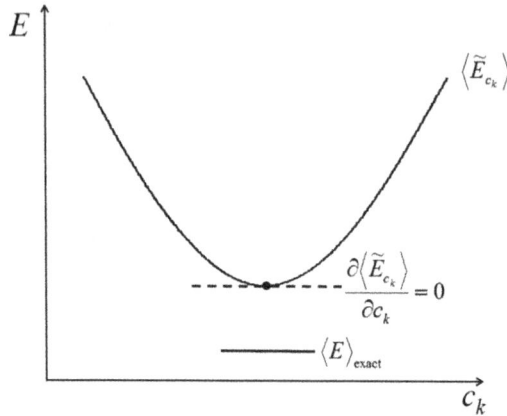

Figure 10.2. Minimizing the energy of trial wavefunction $|\tilde{\Phi}\rangle$ with respect to expansion coefficient c_k according to the variational principle.

Equation (10.23) is rearranged and simplified by inserting equation (10.21) back into equation (10.23) and dividing out common terms. With a little effort the resulting equation is:

$$\sum_{i=1}^{n} c_i \langle \psi_i | \hat{H} | \psi_k \rangle = \sum_{i=1}^{n} c_i \langle \psi_i | \psi_k \rangle E_{c_k} \tag{10.24}$$

The result in equation (10.24) must be repeated to optimize all remaining expansion coefficients. This results in a system of n equations, one for each individual coefficient, to be simultaneously solved. When this system is cast in matrix form, matrix elements differ in location by their indexing coefficient k. As was found to be advantageous in section 7.4, the complete system of equations is conducive to the techniques of linear algebra, and is concisely expressed in matrix form:

$$\mathbf{H\,C} = \mathbf{S\,C\,E} \tag{10.25}$$

Equation (10.25) is called the *secular equation*, in deference to its similarity in form to equations derived for planetary motion. The adjective 'secular' has come to take the meaning of describing the slow changes in relative motion between the planets and the Sun, but the descriptor also carries non-religious connotations, perhaps in reference to application of such equations to planets revolving about the Sun.

The Hamiltonian matrix in equation (10.25) is an $n \times n$ array with elements of the form: $(\mathbf{H})_{ij} = \langle \psi_i | \hat{H} | \psi_j \rangle$. It is symmetric for real basis functions, but in general is Hermitian. The $\mathbf{S}$ matrix also has dimension $n \times n$, and has been previously defined in section 7.2 as the overlap matrix: $(\mathbf{S})_{ij} = \langle \psi_i | \psi_j \rangle$. It is also symmetric for real functions and Hermitian for a complex basis. As defined in section 7.4, $\mathbf{C}$ represents the coefficients, but now instead of being an $n \times 1$ column, simultaneous optimization of all coefficients makes it a square matrix of dimension n. We are hence concurrently finding n of these $n \times 1$ column matrices, each of which represents a unique solution. There is a hierarchy of optimized energy states, with the variational principle assuring all are upper bounds. Equation (10.25) also contains the eigenvalue matrix: $\mathbf{E}$. It is of the same dimension as the others, but is in diagonal form. As is true of any diagonal matrix, it commutes with the $\mathbf{S}$ and $\mathbf{C}$ matrices, so can be placed anywhere on the right-hand side of the equation.

Once $\mathbf{H}$ and $\mathbf{S}$ are constructed from a set of basis functions chosen to expand the trial wavefunction, the task is determining eigenvalue matrix: $\mathbf{E}$ and expansion coefficients: $\mathbf{C}$. One approach is to rearrange the left-hand side so that the right-hand side is the $n \times n$ null matrix:

$$(\mathbf{H} - \mathbf{ES})\mathbf{C} = \mathbf{0} \qquad (10.26)$$

If both sides of equation (10.26) are left multiplied by the matrix: $(\mathbf{H}-\mathbf{ES})^{-1}$, we obtain the unsatisfactory trivial solution: $\mathbf{C} = \mathbf{0}$. This is not possible if the inverse of $(\mathbf{H}-\mathbf{ES})$ does not exist. From linear algebra, a matrix has no inverse under the condition that it has a zero determinant. Hence the energy eigenvalue matrix $\mathbf{E}$ can be found from the known $\mathbf{H}$ and $\mathbf{S}$ matrices by solving the problem:

$$\det(\mathbf{H} - \mathbf{ES}) = 0 \qquad (10.27)$$

Equation (10.25) is more commonly solved by matrix diagonalization. In this fashion, both the eigenvalue and expansion coefficient matrices are simultaneously determined. Suppose as a first example that the basis set is orthonormal. The overlap matrix therefore takes the simplifying form: $\mathbf{S} = \mathbf{1}$ so that equation (10.25) becomes:

$$\mathbf{H}\,\mathbf{C} = \mathbf{C}\,\mathbf{E} \qquad (10.28)$$

Left multiplying equation (10.28) by the inverse of matrix $\mathbf{C}$ results in:

$$\mathbf{C}^{-1}\mathbf{H}\,\mathbf{C} = \mathbf{C}^{-1}\mathbf{C}\,\mathbf{E} = \mathbf{E} \qquad (10.29)$$

Because $\mathbf{E}$ is a diagonal matrix of the energy eigenvalues, the matrix product on the left-hand side of equation (10.29) diagonalizes $\mathbf{H}$. The basis functions we chose for this problem form a real, symmetric Hamiltonian matrix. According to linear algebra, there exists an *orthogonal transformation* which places a symmetric matrix in diagonal form. In addition, an orthogonal transformation matrix obeys the condition:

$$\mathbf{C}^{-1} = \mathbf{C}^{\mathrm{T}} \qquad (10.30)$$

In the general case of a Hermitian Hamiltonian matrix: $\mathbf{H} = \mathbf{H}^{\dagger}$ linear algebra says that it is diagonalizable by a *unitary transformation matrix* with the property: $\mathbf{C}^{-1} = \mathbf{C}^{\dagger}$.

Equation (10.30) explains the implication of its name. Interchange of rows and columns of $\mathbf{C}$ forms its matrix inverse, which in fact shows that the variational expansion wavefunctions are linearly independent. Each of the n columns of $\mathbf{C}$ not only represents a set of weighting factors to the n elements of the basis set, but also collectively form a wavefunction that is both normalized *and* orthogonal to those expanded from the other $n-1$ columns. As a matter of fact, equation (10.30) is a generalization of the condition we stated for our single row vector: $\mathbf{C}^T$ and single column vector $\mathbf{C}$ that was used in our expansion wavefunction of chapter 7, section 7.4.

Suppose that elements of the basis set eigenvector are not orthorgonal, meaning the overlap matrix: $\mathbf{S} \neq \mathbf{1}$ and equation (10.25) must be solved with no simplifications. To accomplish this, we first make use of the following transformations:

$$\mathbf{H'} = \mathbf{S}^{-1/2}\,\mathbf{H}\,\mathbf{S}^{-1/2}$$
$$\mathbf{C'} = \mathbf{S}^{1/2}\,\mathbf{C} \tag{10.31}$$

These definitions allow equation (10.25) to be rewritten in the form:

$$\mathbf{H'}\,\mathbf{C'} = \mathbf{C'}\,\mathbf{E} \tag{10.32}$$

PARALLEL INVESTIGATION: Verify that equation (10.25) is the result of inserting equation (10.24) into equation (10.18).

Just as the original, the transformed Hamiltonian matrix of equation (10.31) also has a symmetric form. As before an orthogonal transformation matrix: $\mathbf{C'}$ results from diagonalizing the reformulated Hamiltonian matrix: $\mathbf{H'}$. If we use the fact that the transpose of a matrix product is: $(\mathbf{A}\,\mathbf{B})^T = \mathbf{B}^T\,\mathbf{A}^T$, it follows in the case of a non-unit overlap matrix that the coefficients and their transformed form have the property:

$$\mathbf{C'^T}\,\mathbf{C'} = \mathbf{C}^T\,(\mathbf{S}^{1/2})^T\,\mathbf{S}^{1/2}\,\mathbf{C} = \mathbf{C}^T\,\mathbf{S}\,\mathbf{C} = \mathbf{1} \tag{10.33}$$

Practical aspects of solving a problem of this type involve forming matrices $\mathbf{S}^{1/2}$ and $\mathbf{S}^{-1/2}$. This is accomplished by first diagonalizing the overlap matrix: $\mathbf{S}$ (which is either symmetric or Hermitian, depending on the chosen basis), taking either the square root or inverse square root of its diagonal elements, then using the orthogonal (or unitary) transformation which diagonalized it to perform a back transformation.

As a variational example, consider the four lowest energy particle in a box solutions defined in chapter 7, equation (7.13), which we will use to find best-fit wavefunctions to the time independent Schrödinger equation including a potential of the form:

$$V(x) = \begin{cases} \infty & x \leqslant 0 \\ 2k(x - L/2)^2 & 0 < x < L \\ \infty & x \geqslant 0 \end{cases} \tag{10.34}$$

At the boundaries, the potential is the same as the original particle in a box problem. Within the box the particle now experiences a potential shown in equation (10.34). The shape of $V(x)$ is plotted in figure 10.3 and is in fact parabolic, reminiscent of the harmonic oscillator potential discussed in chapters 3 and 9. Recall from there the

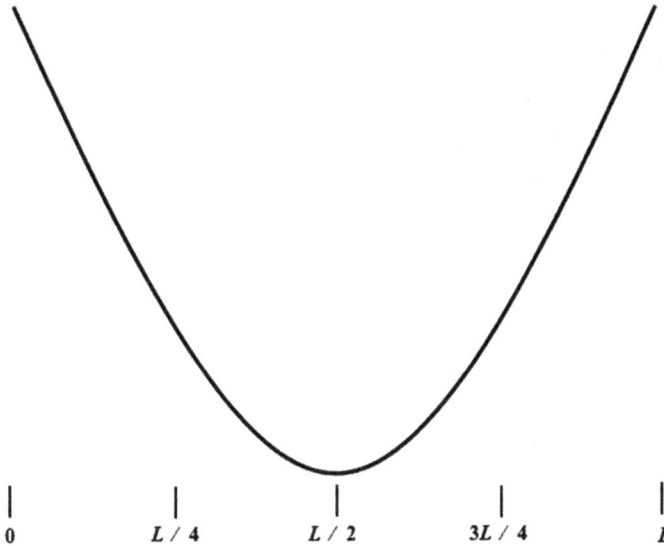

Figure 10.3. Potential $V(x) = 2k(x - L/2)^2$ used for variational treatment of particle in a box solutions as approximations to the harmonic oscillator.

parameter k is the spring constant (SI units: kg $\cdot$ s^{-2}) which defines the potential's steepness. From equation (10.34) we note at the center of the box: $V(L/2) = 0$ and as the the box edges are approached: $V(x \to 0) \to V(x \to L) \to 1/2kL^2$. In chapter 9 we learned the quantum oscillator has analytic wavefunctions of the form presented in equation (9.3), and an exact energy given in equation (10.14).

First consider the lowest energy solution: $|\psi_1\rangle$ within the box. It now not only possesses kinetic energy but is also subject to the modified potential of equation (10.34). In this exercise we will express all matrix elements in atomic unit (see appendix A). The values $\hbar$, m, and L ($=a_0$) are therefore set equal to 1 au. For reasons that come to light later, we choose a spring constant of: $k = 200$ au, which corresponds to an unrealistic SI value of 3.13×10^5 kg $\cdot$ s^{-1}. Using this in the potential term of our model Hamiltonian, the lowest energy particle in a box wavefunction then has energy expectation value:

$$\langle \tilde{E} \rangle = -\frac{1}{2} \cdot 2 \cdot \int_0^1 \sin(\pi x) \frac{d^2}{dx^2} \sin(\pi x) dx + 2 \cdot 400 \cdot \int_0^1 (x - 1/2)^2 \sin^2(\pi x) dx \quad (10.35)$$
$$= 4.9348 + 13.0691 = 18.0039 \text{ au}$$

In comparison, the ground state energy of the quantum mechanical harmonic oscillator wavefunction with this particular force constant is (in atomic units): $\langle E \rangle_{\text{exact}} = (0 + 1/2)\sqrt{k} = 7.07106$ au. It is obvious that our trial wavefunction is not a very good approximation to the actual solution, so let us attempt an expansion of particle in a box wavefunctions to make improvements.

PARALLEL INVESTIGATION: Verify the energy for the trial wavefunction (all quantities in atomic units): $|\tilde{\chi}\rangle = \cos\left(\dfrac{\pi x}{2}\right)$ evaluated using the harmonic oscillator Hamiltonian: $\hat{H} = -\dfrac{1}{2}\dfrac{d^2}{dx^2} + \dfrac{1}{2}kx^2$ and the force constant value: $k = 200$ au is $\langle\tilde{E}\rangle = 14.3028$ au over the range: $-1 \leqslant x \leqslant 1$.

Instead of using only the lowest energy particle in a box wavefunction, suppose we simultaneously find four approximate solutions: $|\Phi_i\rangle$ to the harmonic oscillator Hamiltonian expanded from a basis of the four lowest-energy particle in a box solutions $|\psi_i\rangle$. The basis functions are orthonormal and the overlap matrix constructed from them is: $\mathbf{S} = \mathbf{1}$, so the simplified secular equations of equation (10.28) are to be solved. Since four basis functions are used, the matrix equations produces four orthonormal expansion wavefunctions with different weightings of the basis functions. To facilitate the discussion we will partition the Hamiltonian matrix into its kinetic and potential components: $\mathbf{H} = \mathbf{T} + \mathbf{V}$. The basis functions are eigenfunctions of the kinetic energy operator. According to the discussion in section 7.4, the $\mathbf{T}$ matrix is therefore diagonal with expectation values given in chapter 7, equation (7.41). The kinetic energy matrix expressed in atomic units is:

$$\mathbf{T} = \begin{bmatrix} 4.9348 & 0 & 0 & 0 \\ 0 & 19.7392 & 0 & 0 \\ 0 & 0 & 44.4132 & 0 \\ 0 & 0 & 0 & 78.9568 \end{bmatrix} \tag{10.36}$$

PARALLEL INVESTIGATION: Verify that the first four orthonormal trial wavefunctions: $|\tilde{\chi}_n\rangle = \cos\left(\dfrac{n\pi x}{2}\right) n = 1, 3, 5, 7$ form kinetic energy matrix (with all quantities in atomic units): $\mathbf{T} = \begin{bmatrix} 1.2337 & 0 & 0 & 0 \\ 0 & 11.1003 & 0 & 0 \\ 0 & 0 & 30.84251 & 0 \\ 0 & 0 & 0 & 60.45133 \end{bmatrix}$ over the range: $-1 \leqslant x \leqslant 1$.

In equation (10.36) note the kinetic energy matrix element $(\mathbf{T})_{44} = 78.9568$ au for $|\psi_4\rangle$ makes the magnitude selected for spring constant $k = 200$ au a bit more understandable. Near the two extremes the box has potential: $V = 1/2 \cdot kL^2 = 100$ au. The potential must reach high enough to make sure all four basis functions are 'trapped' within the well. As an additional argument, note that, as discussed in chapter 9, the domain of the exact harmonic oscillator wavefunction is in actuality: $-\infty \leqslant x \leqslant +\infty$. Recall that our particle in a box well is at a maximum at: $L = 0$, falls to zero at: $L = a_0/2$, then is designed to parabolically approach the box wall at: a_0. We are therefore expecting the oscillator to be within the region of space: $-a_0/2 \leqslant x \leqslant +a_0/2$. Using a diatomic molecular force constant such as that for Cl_2 of $k = 329$ kg·s^{-2} ($= 0.211$ au), we can integrate the exact solution probability density to find the likelihood of being within this range. For the exact ground state harmonic oscillator we obtain a 36.2% chance of this localization, however with the steepness changed to 200 au the probability correspondingly increases to a 99.2% chance.

Elements of the potential energy matrix have the form:

$$(V)_{mn} = \frac{2}{L}2k \int_0^L \sin\left(\frac{n\pi x}{L}\right)(x - L/2)^2 \sin\left(\frac{m\pi x}{L}\right)dx \qquad (10.37)$$

Again, these values will be represented in atomic units. With the help of mathematical software the potential energy matrix is:

$$V = \begin{bmatrix} 16.3363 & 0 & 18.9977 & 0 \\ 0 & 35.3341 & 0 & 22.5158 \\ 18.9977 & 0 & 38.8522 & 0 \\ 0 & 22.5158 & 0 & 40.0835 \end{bmatrix} \qquad (10.38)$$

These are combined with equation (10.36) to form the Hamiltonian matrix:

$$H = \begin{bmatrix} 21.2712 & 0 & 18.9977 & 0 \\ 0 & 55.0733 & 0 & 22.5158 \\ 18.9977 & 0 & 83.2654 & 0 \\ 0 & 22.5158 & 0 & 119.0403 \end{bmatrix} \qquad (10.39)$$

PARALLEL INVESTIGATION: Verify that the first four orthonormal trial wave-functions: $|\bar{\chi}_n\rangle = \cos\left(\frac{n\pi x}{2}\right) n = 1, 3, 5, 7$ form a Hamiltonian matrix (with all quantities in atomic units): $H = \begin{bmatrix} 14.3028 & -15.19818 & 2.81448 & -0.98507 \\ -15.19818 & 42.18506 & -18.99772 & 4.25549 \\ 2.81448 & -18.99772 & 63.36528 & -19.70134 \\ -0.98507 & 4.25549 & -19.70134 & 93.3711 \end{bmatrix}$ over the range:

$-1 \leqslant x \leqslant 1$. When forming the integrals, use the actual harmonic oscillator potential from equation (9.1) and force constant $k = 200$ au.

The Hamiltonian matrix in equation (10.39) is diagonalized using mathematical software, giving the following eigenvalues:

$$E = \begin{bmatrix} 15.9126 \\ 47.9428 \\ 88.6240 \\ 126.1709 \end{bmatrix} \qquad (10.40)$$

The expansion coefficients for the four particle in a box wavefunctions are contained in the columns of the orthogonal transformation which diagonalizes the Hamiltonian matrix:

$$C = \begin{bmatrix} 0.9624 & 0 & 0.2715 & 0 \\ 0 & 0.9533 & 0 & 0.3019 \\ -0.2715 & 0 & 0.9624 & 0 \\ 0 & -0.3019 & 0 & 0.9533 \end{bmatrix}$$

(10.41)

Note that the sum of the square of elements in any column is unity, or that each is normalized. In addition, the dot product of any column with another is zero, showing the vectors are orthogonal. The orthonormality of the expansion wavefunctions is succinctly demonstrated via the matrix product: $C^T C = 1$. Inspecting the Hamiltonian matrix in equation (10.39), we see values of zero for the 12 (or 21) and 14 (or 41) elements. Since there is no coupling of basis function $|\psi_1\rangle$ to either basis function $|\psi_2\rangle$ or $|\psi_4\rangle$ across the Hamiltonian, they do not mix together in any of the expansion wavefunctions. Likewise, the lack of coupling in the Hamiltonian matrix between basis function $|\psi_2\rangle$ with $|\psi_1\rangle$ or $|\psi_3\rangle$ precludes their interaction in any of the $|\Phi_i\rangle$.

Employing an expansion of four particle in a box wavefunctions instead of a single function lowered the ground state trial wavefunction energy from to 18.000 39 to 15.9126 au. When compared to the exact ground state energy: $E_0 = 7.071\,106$ au, this is an improvement, but is still nothing to write home about. However, if you have been following the parallel investigations for trial wavefunctions expanded from a basis of cosine functions evaluated over the actual harmonic oscillator Hamiltonian, we find the very satisfactory result presented below.

PARALLEL INVESTIGATION: Verify the Hamiltonian matrix from the previous parallel investigation produces energy eigenvalues: $E = \begin{bmatrix} 7.07119 \\ 35.39218 \\ 64.77328 \\ 105.98758 \end{bmatrix}$ for the first four orthonormal trial wavefunctions: $|\tilde{\chi}_n\rangle = \cos\left(\dfrac{n\pi x}{2}\right)$ $n = 1, 3, 5, 7$ over the range: $-1 \leqslant x \leqslant 1$. The ground state energy of this trial wavefunction is: 0.000 13 au or 0.0018% above the exact ground state energy! As a matter of fact the other three eigenvalues represent very reasonable upper bounds to exact harmonic oscillator states: $E_2 = 35.355\,34$ au (in error by: 0.104%), $E_4 = 63.639\,61$ au (1.78%) and $E_6 = 91.923\,88$ au (15.30%). The expansion wavefunctions are shown in figure 10.4, and should be compared to the exact harmonic oscillator wavefuntions in figure 9.1.

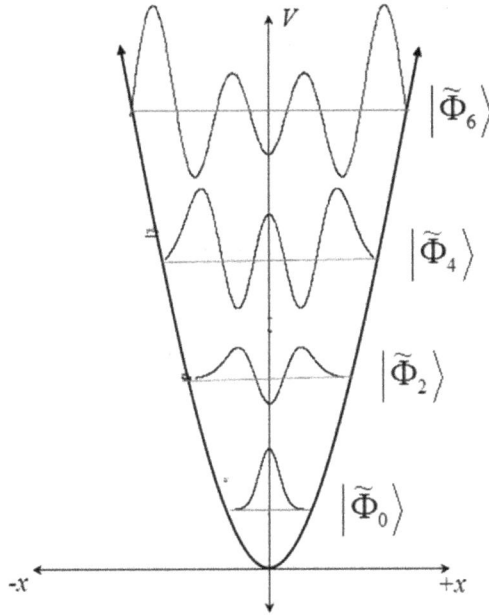

Figure 10.4. The variational wavefunctions found as approximate solutions to the harmonic oscillator. The basis function expansion coefficients are taken from the orthogonal transformation matrix which diagonalizes the Hamiltonian:

$$|\tilde{\Phi}_0\rangle = 0.890|\tilde{\chi}_1\rangle + 0.443|\tilde{\chi}_3\rangle + 0.110|\tilde{\chi}_5\rangle + 0.013|\tilde{\chi}_7\rangle$$

$$|\tilde{\Phi}_2\rangle = -0.410|\tilde{\chi}_1\rangle + 0.670|\tilde{\chi}_3\rangle + 0.601|\tilde{\chi}_5\rangle + 0.148|\tilde{\chi}_7\rangle$$

$$|\tilde{\Phi}_4\rangle = 0.192|\tilde{\chi}_1\rangle - 0.555|\tilde{\chi}_3\rangle + 0.622|\tilde{\chi}_5\rangle + 0.518|\tilde{\chi}_7\rangle$$

$$|\tilde{\Phi}_6\rangle = .060|\tilde{\chi}_1\rangle - 0.216|\tilde{\chi}_3\rangle + 0.490|\tilde{\chi}_5\rangle - 0.842|\tilde{\chi}_7\rangle$$

What's the Matter with Waves?
An introduction to techniques and applications of quantum mechanics
William Parkinson

Chapter 11

Electrons in atoms

11.1 Rotational motion due to a central potential: the hydrogen atom

Both chapter 4 and chapter 8 employed a rigid rotor constraint, fixing radial component r to describe rotational motion through angular variation. The resulting models were expressed by tractable equations. Comparing equations (8.15) and (8.26) shows how the kinetic energy terms for rotation become more unruly as we go from the circular to the spherical case, as the number of independent variables increase from only ϕ to both θ and ϕ. Transitioning to atoms requires a completely general model of spherical motion with three independent degrees of freedom, whether expressed in the Cartesian x, y, z or spherical polar r, θ, ϕ frame. Intuition suggests we again take the spherical polar approach for electrons in atoms, but it should be anticipated that increasing the particle's degrees of freedom will again increase the complexity of kinetic energy operator: $\hat{T}$. To further exacerbate the situation, modeling a charged particle like the electron moving about a charged nucleus now requires we also include a radially-dependent attractive electrostatic potential $\hat{V}$.

Recall the strategy employed in chapter 8. Kinetic energy of rotating particles was introduced through angular momentum using: $\hat{T} = \hat{L}^2/2I$. For atoms, it is advantageous to return to the kinetic energy operator: $\hat{T} = -\hbar^2\hat{\nabla}^2/2\mu$, and perform a series of partial derivative conversions to transform variables of the del operator from Cartesian to spherical polar form. Beginning with the definitions in chapter 8, equation (8.2), the manipulations proceed in a very similar fashion to what was done in section 8.1. This is left as an exercise to the motivated reader. After a few pages of algebra, the differential operator for kinetic energy is:

$$\hat{T} = -\frac{\hbar^2}{2\mu}\hat{\nabla}^2 = -\frac{\hbar^2}{2\mu}\left[\frac{\partial^2}{\partial x^2} + \frac{\partial^2}{\partial y^2} + \frac{\partial^2}{\partial z^2}\right] = -\frac{\hbar^2}{2\mu}\left[\frac{\partial^2}{\partial r^2} + \frac{2}{r}\frac{\partial}{\partial r}\right] + \frac{1}{2\mu r^2}\hat{L}^2 \quad (11.1)$$

The angular momentum operator: $\hat{L}^2$ is defined in chapter 8, equation (8.26). Comparing rotational kinetic energy operators from chapter 8, equations (8.15)

doi:10.1088/978-1-6817-4577-0ch11 11-1

and (8.26) to equation (11.1), we note the hierarchy of complexity as expressions evolve from a function of the single variable ϕ to a complete function of variables r, θ, and ϕ.

During the classical treatment of chapter 4, it was stated that a body undergoes rotational motion only when acted upon by a central force. We could therefore induce an electron into spherical motion by tethering it, or possibly placing a massive body at the origin causing gravitational attraction. Of course a negatively charged particle moves under the influencing force of Coulombic attraction to the nucleus. Electrostatics described by a radial potential used in chapter 4, equation (4.29) now quantum mechanically takes operator form:

$$\hat{V}(r) = -\frac{Ze^2}{4\pi\varepsilon_0 \hat{r}} \tag{11.2}$$

This has physical constants defined previously, with Z being the atomic number of the nucleus and the inverse of the radial position operator: $\hat{r} = r\times$. Application of the operator in this case requires division instead of multiplication. The sign of $\hat{V}(r)$ reflects the opposite magnitude of elementary charge for protons and electrons.

We seek eigenfunctions of the time independent Schrödinger equation: $\hat{H}\psi(r, \theta, \phi) = E\psi(r, \theta, \phi)$ with Hamiltonian: $\hat{H} = \hat{T} + \hat{V}$ defined by equations (11.1) and (11.2). As has proved successful time and again we try separation of variables, with a product eigenfunction taking the form:

$$\psi(r, \theta, \phi) = R_n(r) \cdot Y_\ell^{m_\ell}(\theta, \phi) \tag{11.3}$$

Angular dependence in equation (11.3) is expressed through a spherical harmonic (see section 8.2 and table 8.1). The product wavefunction is formed by multiplying $Y_\ell^{m_\ell}(\theta, \phi)$ by a so-called *radial function*: $R_n(r)$.

Justifying variable separation in equation (11.3) requires arguments from section 7.5 involving commuting operators and simultaneously measurable properties. Spherical harmonics are eigenfunctions of $\hat{L}^2$ with dependence on polar angle θ and azimuthal angle ϕ. Examining the commutation relation between: $\hat{L}^2$ and the hydrogen atom Hamiltonian: $\hat{H} = \hat{T} + \hat{V}$ we find:

$$[\hat{H}, \hat{L}^2] = -\frac{\hbar^2}{2\mu}\left(\left[\frac{\partial^2}{\partial r^2}, \hat{L}^2\right] + \left[\frac{2}{r}\frac{\partial}{\partial r}, \hat{L}^2\right]\right) + \frac{1}{2\mu r^2}[\hat{L}^2, \hat{L}^2] + [\hat{V}(r), \hat{L}^2] = 0 \quad (11.4)$$

Recall from chapter 8, equation (8.26) that $\hat{L}^2$ involves differentiation with respect to θ and ϕ. The two leftmost terms of the right-hand side of equation (11.4) are therefore zero. The third term involves an operator commuting with itself, which is always true. Finally, the radial potential has no angular dependence, so the last commutator of the right-hand side of equation (11.4) is also zero.

The operator for the z-component of angular momentum, $\hat{L}_z$ from chapter 8, equation (8.15), is solely a function of azimuthal angle ϕ. It was also shown in section 8.2 that: $[\hat{L}^2, \hat{L}_z] = 0$. Similar to equation (11.4), it is straightforward to show that: $[\hat{H}, \hat{L}_z] = 0$ as well. As a consequence of these facts, it will be possible to

Table 11.1. Radial wavefunctions for $n = 1$ to $n = 3$.

n	ℓ	$R_n(r)$ [11]
1	0	$2\left(\dfrac{Z}{a_0}\right)^{3/2} \cdot e^{-Zr/a_0}$
2	0	$\dfrac{1}{\sqrt{2}}\left(\dfrac{Z}{a_0}\right)^{3/2}\left(1 - \dfrac{Zr}{2a_0}\right) \cdot e^{-Zr/2a_0}$
2	1	$\dfrac{1}{2\sqrt{6}}\left(\dfrac{Z}{a_0}\right)^{5/2} \cdot r e^{-Zr/2a_0}$
3	0	$\dfrac{2}{3\sqrt{3}}\left(\dfrac{Z}{a_0}\right)^{3/2}\left(1 - \dfrac{2Zr}{3a_0} + \dfrac{2Z^2r^2}{27a_0^2}\right) \cdot e^{-Zr/3a_0}$
3	1	$\dfrac{8}{27\sqrt{6}}\left(\dfrac{Z}{a_0}\right)^{3/2}\left(\dfrac{Zr}{a_0} - \dfrac{Z^2r^2}{6a_0^2}\right) \cdot e^{-Zr/3a_0}$
3	2	$\dfrac{4}{81\sqrt{30}}\left(\dfrac{Z}{a_0}\right)^{7/2} \cdot r^2 e^{-Zr/3a_0}$

[1] $a_0 = \dfrac{(4\pi\epsilon_0)\hbar^2}{m_e e^2} = 0.529\text{Å}$

construct solutions to the central potential problem which are simultaneous eigenfunctions of $\hat{H}$, $\hat{L}^2$, and $\hat{L}_z$.

Equation (11.3) is inserted into the time independent Schrödinger equation using a Hamiltonian containing both equations (11.1) and (11.2). We then use the fact that the spherical harmonics are eigenfunctions of $\hat{L}^2$ to write:

$$\hat{H}\psi(r,\,\theta,\,\phi) = \left[-\frac{\hbar^2}{2\mu}\left(\frac{\partial^2}{\partial r^2} + \frac{2}{r}\frac{\partial}{\partial r}\right) + \frac{1}{2\mu r^2}\hat{L}^2 + V(r)\right]R_n(r) \cdot Y_\ell^{m_\ell}$$

$$= \left[-\frac{\hbar^2}{2\mu}\left(\frac{\partial^2}{\partial r^2} + \frac{2}{r}\frac{\partial}{\partial r}\right) + \frac{\ell(\ell+1)\hbar^2}{2\mu r^2} + V(r)\right] \tag{11.5}$$

$$\times R_n(r) \cdot Y_\ell^{m_\ell} = ER_n(r) \cdot Y_\ell^{m_\ell}$$

The second and third lines of equation (11.5) are divided through by $Y_\ell^{m_\ell}(\theta,\,\phi)$, giving a differential equation in terms of the radial function:

$$\left[-\frac{\hbar^2}{2\mu}\left(\frac{\partial^2}{\partial r^2} + \frac{2}{r}\frac{\partial}{\partial r}\right) + \frac{\ell(\ell+1)\hbar^2}{2\mu r^2} + V(r)\right]R_n(r) = ER_n(r) \tag{11.6}$$

When this is rearranged and expressed in atomic units (assuming $m_e = \mu$, see appendix A) and primes are used as indicators of derivatives with respect to r, the resulting linear homogeneous second order differential equation is:

$$R_n''(r) + \frac{2}{r}R_n'(r) + \left[2E + \frac{2Z}{r} - \frac{\ell(\ell+1)}{r^2}\right]R_n(r) = 0 \tag{11.7}$$

We skip the rigors of solving this equation which, for those interested in mathematical physics, can be found in a variety of sources. The $R_n(r)$ have a mathematical form of exponentials multiplied by Laguerre polynomials. The first several radial wavefunctions are presented in table 11.1. These are expressed using the Bohr radius, a_0, a quantity first encountered in chapter 4, equation (4.27). Multiplying the radial functions by an appropriate spherical harmonic gives the product wavefunctions of equation (11.3) which are simultaneous eigenfunctions of: $\hat{H}$, $\hat{L}^2$, and $\hat{L}_z$.

11.2 Properties of the hydrogen atom solutions

Recalling from chapter 8, quantum numbers ℓ and m_ℓ arose in the spherical harmonics as a consequence of the particle on a ring and particle on a sphere solutions. Boundary conditions of the radial equations introduce a new dependence requiring an additional quantum number:

$$n = 1, 2, 3, \ldots \tag{11.8}$$

Eigenvalues of the radial equation show how the bound state energy levels of a hydrogen or hydrogen-like atom depend on this quantum number:

$$E_n = -\frac{1}{2}\frac{Z^2}{n^2}\frac{\mu e^4}{\hbar^2(4\pi\varepsilon_0)^2} \tag{11.9}$$

Using the definitions of the Bohr radius or Hartree energy unit, equation (11.9) is written:

$$E_n = -\frac{1}{2}\frac{Z^2}{n^2}\frac{\hbar^2}{\mu a_0^2} = -\frac{1}{2}\frac{Z^2}{n^2}E_h \tag{11.10}$$

With the time independent Schrödinger equation written in atomic units and $R_1(r)$ expressed similarly, we see the lowest energy $Z = 1$ radial wavefunction is a satisfactory eigenfunction returning the appropriate eigenvalue:

$$\hat{H}R_1(r) = \left[-\frac{1}{2}\frac{\partial^2}{\partial r^2} - \frac{1}{r}\frac{\partial}{\partial r} - \frac{1}{r}\right]2 \cdot e^{-r}$$
$$= \left[-\frac{1}{2} + \frac{1}{r} - \frac{1}{r}\right]2 \cdot e^{-r} = \left(-\frac{1}{2}\mathrm{au}\right) \cdot R_1(r) \tag{11.11}$$

First it is noted that the energy expressions of equations (11.9) or (11.10) are the exact same result obtained by Bohr (see chapter 4) some two decades prior to the general quantum mechanical formulation. His derivation modified a classically rotating electron to the constraint that that its angular momentum values must be integer multiples of $\hbar$. Recall that Bohr's result was limited to one electron atoms. In principle, the Hamiltonian of the Schrödinger equation can include terms that make it generally applicable to a multi-particle system of any composition.

The next point of importance is that the bound state energy levels are discrete functions of the positive integer quantum number n. Because only the radial components have n-dependence, the energy is therefore determined solely from the radial wavefunction. However, as seen in table 11.1, the radial wavefunctions themselves have parametric dependence on angular quantum number ℓ, which becomes apparent upon inspection of equation (11.7). The probability density is formed from the square of equation (11.3). Its integration over all space requires a triple integral with the spherical polar volume element:

$$\int d\tau = \int_0^{2\pi} d\phi \cdot \int_0^{\pi} \sin(\theta)d\theta \cdot \int_0^{\infty} r^2 dr \qquad (11.12)$$

Expectation values involving the radial functions given in table 11.1 are evaluated according to the recipe:

$$\langle R_n(r) | \hat{O}(r) | R_n(r) \rangle = \int_0^{\infty} r^2\, dr \cdot R_n(r)\hat{O}(r)R_n(r) \qquad (11.13)$$

As an example, we take $R_2(r)$ for hydrogen ($Z = 1$) and the Hamiltonian, both expressed in atomic units, then use integrals from section B.2 to find:

$$
\begin{aligned}
\langle R_2(r) | \hat{H} | R_2(r) \rangle &= \frac{1}{2} \cdot \int_0^{\infty} r^2 dr \cdot (1 - r/2)e^{-r/2} \\
&\quad \times \left[-\frac{1}{2}\frac{\partial^2}{\partial r^2} - \frac{1}{r}\frac{\partial}{\partial r} - \frac{1}{r} \right](2 - r/2)e^{-r/2} \\
&= \frac{1}{2} \cdot \int_0^{\infty} \left[-\frac{7r^2}{8} + \frac{3r^3}{8} - \frac{r^4}{32} \right]e^{-r}dr = -0.125 \text{ au}
\end{aligned}
\qquad (11.14)
$$

PARALLEL INVESTIGATION: Verify that: (a) The radial wavefunctions are normalized by evaluating: $\langle R_1(r)|R_1(r)\rangle$. (b) The radial wavefunctions are orthogonal by evaluating: $\langle R_1(r)|R_2(r)\rangle$. (c) The energy expectation value: $\langle R_3(r)|\hat{H}|R_3(r)\rangle = -0.0556$ au.

The results of equations (11.11) or (11.14) are values for *bound state* or negative energy levels, being measured relative to the zero point of an infinitely-separated proton and electron. The parametric role that ℓ plays in the form of $R_n(r)$ is verified in table 11.1, which shows the radial functions take varying forms depending upon their individual ℓ value. Introduction of n imposes a boundary condition that modifies allowed values of ℓ compared to a particle on a sphere in section 8.2. For a hydrogen atom, the allowed values of ℓ are now:

$$\ell = 0, 1, 2, ..., n - 1 \qquad (11.15)$$

Hydrogen energy levels exhibit a $2\ell + 1$ degeneracy, as do the particle on a sphere energy levels depicted in chapter 8, figure 8.2. The values allowed for m_ℓ remain those given in chapter 8, equation (8.28).

The quantum number n is known as the *principal quantum number*, not only because it controls allowed values for ℓ and energy of a level (see equation (11.10)), but it also parametrically determines the average radial distance of an electron through $R_n(r)$. The value ℓ, the *angular quantum number*, determines angular characteristics including electron angular momentum of magnitude: $\sqrt{\ell(\ell + 1)}\,\hbar$. Spherical harmonics of a given ℓ value are referred to as *orbitals*. In fact, the angular quantum number is also referred to as the orbital quantum number. It is common to replace the numeric value of ℓ by its more familiar letter designation:

$$
\begin{array}{c|c|c|c|c|c|c|c|c|c|c}
 & 0 & 1 & 2 & 3 & 4 & 5 & 6 & 7 & 8 & 9 \\
\hline
\text{letter} & s & p & d & f & g & h & i & k & l & m
\end{array}
\tag{11.16}
$$

The first four orbital letter designations pay tribute to the four Balmer series visible emission spectral lines of the hydrogen atom. The letters represent the original German names of the spectral lines, but the English adaptations amount to a transition characterized by a narrow or 'sharp' line, another of highest intensity designated the 'primary' line, a third appearing somewhat fuzzy or 'diffuse', and one referencing the 'fundamental'. Following the first four orbital descriptors, those of sequential ℓ take letter designations in alphabetic order beginning with g (but omitting j).

Quantum number n sequentially labels each energy level, and as equation (11.15) shows, also allows the introduction of one new ℓ value per level. Hence each energy level introduces an additional available orbital of increased angular momentum proportional to: $\sqrt{\ell(\ell + 1)}$. As a result, every energy level: $1 \to \infty$ possesses an s-type orbital of zero angular momentum. Levels: $2 \to \infty$ include a p-type orbital with angular momentum: $L = \sqrt{2}\,\hbar$, Levels: $3 \to \infty$ have a d-type type of angular momentum: $L = \sqrt{6}\,\hbar$, etc. To differentiate between the various orbitals of the same angular momentum, the principal quantum number is listed before the letter designation such as: 1s, 2s, 3s, ...

As discussed for a particle on a ring in section 8.1, m_ℓ measures the z-projection of angular momentum in units: $L_z = m_\ell \hbar$. The name given to m_ℓ stems from the effect of a static magnetic field on the magnetic moment of an electron, which causes splits in degeneracy of orbitals of a given ℓ value. Atomic spectroscopy performed under this condition, known as the Zeeman effect, will exhibit a fine structure caused by the slight difference in energy levels due to orbitals lowered, raised, or unchanged in energy upon interaction with the magnetic field.

Figure 11.1 displays amplitude plots of selected radial functions of hydrogen ($Z = 1$). The first three *s*-type functions are displayed in figure 11.1(a)–(c). For clarity, each has its own appropriate abscissa scaled in integer multiples of the Bohr radius $a_0 = 0.529$ Å. In addition to an asymptotic node, these show the hierarchy of $n-1$ radial nodes per

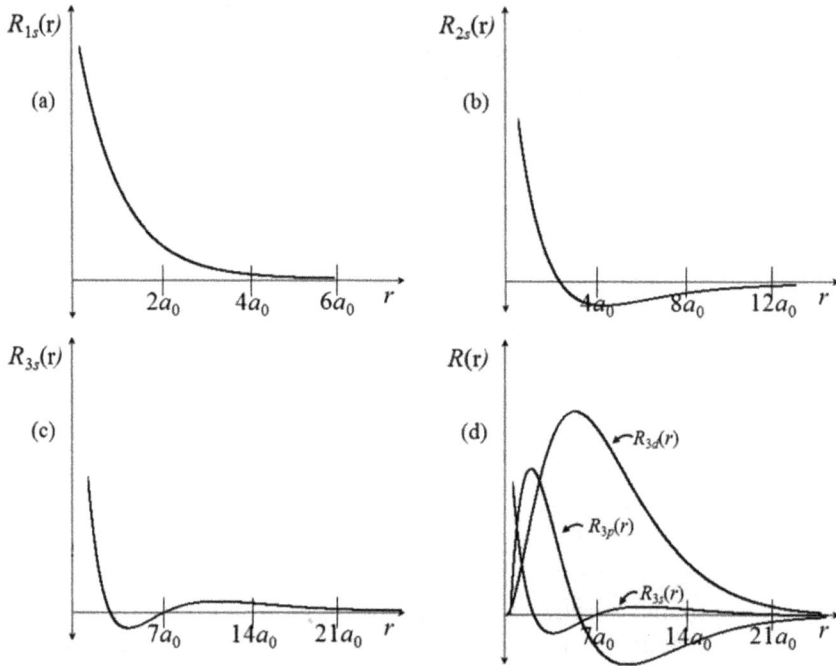

Figure 11.1. Plots of various radial wavefunctions for hydrogen ($Z = 1$). (a) The 1s function. (b) The 2s function. (c) The 3s function. (d) Plots of the radial wavefunctions for the three allowed angular momentum values of $n = 3$.

function. Referring to table 11.1, $R_{1s}(r)$ has indeed has exponential form. The radial node of $R_{2s}(r)$ can be determined by setting its Laguerre polynomial to zero:

$$1 - \frac{r}{2a_0} = 0 \qquad (11.17)$$

which occurs at $r = 2a_0$.

PARALLEL INVESTIGATION: Verify that $R_{3s}(r)$ has radial nodes at: $r = 1.9a_0$ and $r = 7.1a_0$.

Figure 11.1(d) shows the radial amplitude plot of the three angular momentum values allowed on energy level $n = 3$. The pattern displayed there is typical of all energy levels. The highest angular momentum function (in this case $R_{3d}(r)$) only possesses an asymptotic node. Radial functions of decreasing angular momentum sequentially add one radial node. In figure 8.3(d), $R_{3p}(r)$ has one and $R_{3s}(r)$ two radial nodes.

The average distance from the nucleus to an electron is not described by its radial amplitude, but as discussed in section 7.1, requires knowledge of its probability density. The symmetry of atomic systems mandates a spherical surface average known as the *radial probability density*. This is constructed by multiplying the wavefunction square modulus by a spherical shell volume element. Because radial functions are real, this quantity is:

$$4\pi r^2 \cdot R_n^2(r) \tag{11.18}$$

Figure 11.2 contains selected radial probability density plots of hydrogen ($Z = 1$). The first three s-type probability shells are displayed in figure 11.2(a)–(c). As in figure 11.1, each has its own appropriate abscissa scaled in integer multiples of the Bohr radius $a_0 = 0.529$ Å. The shape of figure 11.2(a) shows the radial probability density reaches a maximum. This function can therefore be used to find the inflection point corresponding to the most probable radial distance: r^*. To do so we use equation (11.18) along with the function $R_{1s}(r)$ of hydrogen ($Z = 1$), and locate the point of zero slope for the radial shell probability density:

$$\frac{d}{dr}(4\pi r^2 \cdot R_{1s}^2(r)) = 0 = 4\pi \cdot 4\frac{1}{a_0^3}\left[2r \cdot e^{-2r/a_0} - \frac{2r^2}{a_0} \cdot e^{-2r/a_0}\right] \tag{11.19}$$

Solving equation (11.19) we find:

$$r^* = a_0 = 0.529 \text{ Å} \tag{11.20}$$

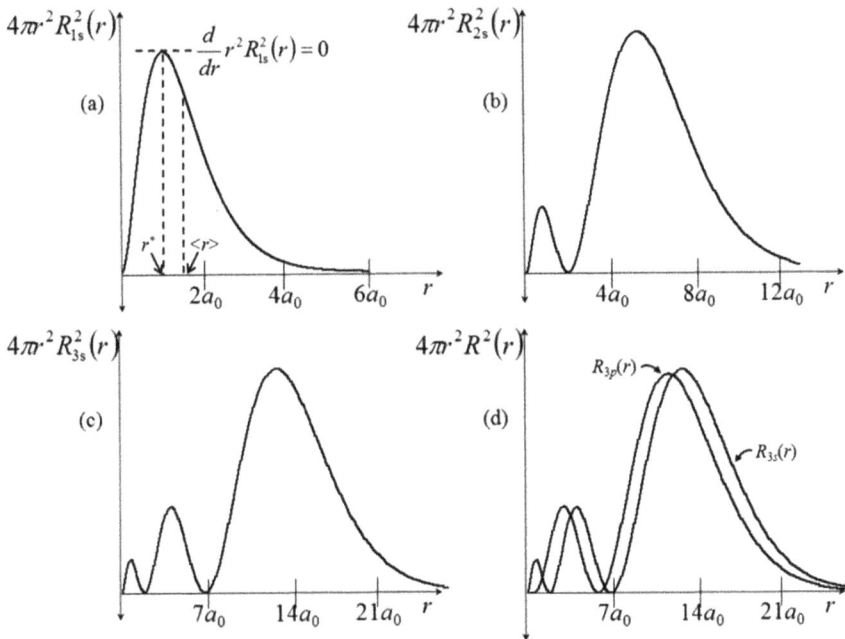

Figure 11.2. Plots of various radial shell probability densities for hydrogen ($Z = 1$). (a) The 1s function. (b) The 2s function. (c) The 3s function. (d) Plots of the radial wavefunctions for the $\ell = 0$ and $\ell = 1$ angular momentum values of $n = 3$.

This of course is the result obtained by Bohr from his one-electron model.

> **PARALLEL INVESTIGATION:** Verify that the $n = 2$, $\ell = 0$ radial function of hydrogen ($Z = 1$) has most probable radial distance: $r^* = 6a_0$.

If instead we determine the average distance to the electron in the hydrogen 1s orbital, we use the expectation value recipe, which is done with the help of integrals in section B.2 applied with the spherical polar radial integration element:

$$\langle r \rangle = \langle R_{1s}(r)|\hat{r}|R_{1s}(r)\rangle = 4\frac{1}{a_0^3}\int_0^\infty r^2\,dr \cdot re^{-2r/a_0} = 1.5a_0 = 0.794\ \text{Å} \qquad (11.21)$$

> **PARALLEL INVESTIGATION:** Verify for the $n = 2$, $\ell = 0$ radial function of hydrogen ($Z = 1$) has average radial distance: $\langle r \rangle = 5a_0$.

The skewed shape of the 1s radial shell probability density results in an electron positioned on average 50% past the Bohr radius. Indicated in figure 11.2(a) are the locations of both r^* and $\langle r \rangle$.

As was demonstrated for the 1-dimensional particle in a box wavefunction in section 7.4, it is also possible to evaluate the probability density over fractions of the radial interval to determine the likelihood of the electron occupying that region of space. For instance, the probability of finding the 1s hydrogen electron somewhere between the nucleus and the Bohr radius is:

$$4\frac{1}{a_0^3}\int_0^{a_0} r^2\,dr \cdot e^{-2r/a_0} = 0.323 \qquad (11.22)$$

which is a 32.3% chance. Notice that equation (11.22) integrates *only* the probability density multiplied by the radial integration element. If the probability density for the 1s electron is instead evaluated between a_0 and ∞, the result is 0.677 or a 67.7% chance, as anticipated.

> **PARALLEL INVESTIGATION:** Verify for the $n = 1$, $\ell = 0$ radial function of hydrogen ($Z = 1$) that its electron has a 43.9% chance of being between a_0 and $2a_0$.

The complete wavefunction is a product of a radial wavefunction and a spherical harmonic which itself is an eigenfunction of both $\hat{L}^2$ and $\hat{L}_z$ with coordinates expressed in the spherical polar frame. It is also possible to convert hydrogen atom solutions into Cartesian form. Using chapter 7, table 7.1 and table 11.1 along with the spherical polar coordinate definitions given in chapter 8, equation (8.2), the $n = 2$, $\ell = 1$, $m_\ell = 0$ function converts to:

$$\psi_{2,1,0} = \frac{1}{4\sqrt{2\pi}}\left(\frac{Z}{a_0}\right)^{5/2} \cdot r\cos(\theta) \cdot e^{-Zr/2a_0}$$

$$= \frac{1}{4\sqrt{2\pi}}\left(\frac{Z}{a_0}\right)^{5/2} \cdot z \cdot e^{-Z(x^2+y^2+z^2)^{1/2}/2a_0} \tag{11.23}$$

Using language familiar to chemistry, this is the $2p_z$ orbital.

Constructing other Cartesian orbitals requires linear combinations of angular momentum functions, which has the added benefit of producing real functions through the elimination of imaginary components. For instance, using tables 8.1 and 11.1 with equation (8.2), we take the combination:

$$\psi_{2p_x} = \frac{1}{\sqrt{2}} \cdot (\psi_{2,1,-1} - \psi_{2,1,+1})$$

$$= \frac{1}{8\sqrt{2\pi}}\left(\frac{Z}{a_0}\right)^{5/2} \cdot r\sin(\theta)[e^{-i\phi} + e^{+i\phi}] \cdot e^{-Zr/2a_0}$$

$$= \frac{1}{4\sqrt{2\pi}}\left(\frac{Z}{a_0}\right)^{5/2} \cdot r\sin(\theta) \cdot \cos(\phi) \cdot e^{-Zr/2a_0} \tag{11.24}$$

$$= \frac{1}{4\sqrt{2\pi}}\left(\frac{Z}{a_0}\right)^{5/2} \cdot x \cdot e^{-Z(x^2+y^2+z^2)^{1/2}/2a_0}$$

The imaginary exponentials of equation (11.24) were eliminated using: $e^{\pm i\phi} = \cos(\phi) \pm i\sin(\phi)$. In a similar fashion we find: $1/i\sqrt{2} \cdot (\psi_{2,1,-1} + \psi_{2,1,+1}) = \psi_{2p_y}$. This can be continued to form a set of five real d functions and seven real f functions. When these are applied as orbital representations in the description of an arbitrary many electron atom, they are referred to as *hydrogenic orbitals*. It should also be pointed out that alternate definitions of spherical harmonics exist in which the functions with odd values of m_ℓ do not possess a phase factor, hence the signs on hydrogenics in the above linear combinations are reversed.

PARALLEL INVESTIGATION: Verify that the linear combination:

$$1/\sqrt{2} \cdot (\psi_{3,2,-2} + \psi_{3,2,+2}) = \psi_{3d_{x^2-y^2}}$$

Any wavefunction formed from a combination of eigenfunctions is itself an eigenfunction of the Hamiltonian, and in fact has the same eigenvalue as the original functions. To see this is the case, suppose we form a new function from a weighted combination of two functions which obey the conditions: $\hat{O}\psi_1 = a\psi_1$ and $\hat{O}\psi_2 = a\psi_2$. It follows that:

$$\hat{O}\Psi = \hat{O}(c_1\psi_1 + c_2\psi_2) = (ac_1\psi_1 + ac_2\psi_2) = a(c_1\psi_1 + c_2\psi_2) = a\Psi \tag{11.25}$$

Therefore, the energy eigenvalue of ψ_{2p_z} is the same as ψ_{2p_x} and ψ_{2p_y}, which are the same as ψ_{2s} for that matter, and have value:

$$E_2 = -\frac{Z^2\mu e^4}{8\hbar^2(4\pi\varepsilon_0)^2} = -\frac{Z^2}{8}E_h \tag{11.26}$$

There is a trade-off to be considered when converting eigenfunctions of the central potential problem to Cartesian form. A majority require taking linear combination of some type to eliminate terms of the form: $e^{\pm im_\ell\phi}$. A benefit is the accompanying elimination of imaginary components to form a real wavefunction. A downside is the resulting functions are no longer eigenfunctions of angular momentum. This is a mathematical consequence, but is also an artifact of uncertainty. Combining functions has altered the variance in average values of two conjugate variables.

11.3 Electron spin

Before moving to many-electron atoms, we must address another intriguing aspect necessitated when describing physical phenomena through quantum mechanics. This unfortunately has no classical analog, thus creates a situation where our ability to make connection with the familiar is challenged. In addition to any angular momentum a rotating particle experiences through the mechanisms we have already discussed, subatomic particles—the protons, neutrons, and electrons which comprise all matter—possess an additional *intrinsic angular momentum*. First proposed by Uhlenbeck and Goudsmit, it is referred to as a particle's *spin*. The name suggests a classical connection to the effect resulting from a charge rotating on its own axis. This motion would produce a magnetic moment, with opposing directions of rotation flipping orientation of its north magnetic pole. Although this in some way rationalizes spin characteristics of protons and electrons, a zero-charge neutron also exhibits this phenomenon.

Particle spin arises naturally in Dirac's development of quantum mechanics. Throughout this book we follow the approach of Schrödinger and what is referred to as *non-relativistic quantum mechanics*. As a result the existence of spin must be postulated. We will further assume that spin and spatial coordinates are separable. In the terminology of section 7.5 this means space and spin operators commute, so our wavefunctions can be simultaneous eigenfunctions to both. This in general is not the case for electrons in an atom, where cross-terms of the orbital and spin angular momentum, familiarly known as *spin–orbit interactions*, make their individual coordinates inseparable. These affects only become significant as the nucleus becomes fairly large, so can be neglected for elements in the first few rows of the periodic table.

In the non-relativistic development, eigenfunctions and eigenvalues of spin emerge from two sources of information. First, we interject results gleaned from *spatial angular momentum*, the quantum mechanical study of orbital angular momentum for a rotating object. Secondly, we apply the results of the 1922 experiment of Stern and Gerlach, which passed a low-intensity beam of silver atoms through an inhomogeneous magnetic field. The beam was split into two bands, suggesting two orientations of intrinsic or spin angular momentum. Applying knowledge gained from section 8.2 spatial angular momentum has the possibility of $2\ell + 1$ orientations, a result of quantum number m_ℓ taking integer values from $-\ell$ to $+\ell$. Analogously, the two bands of the Stern–Gerlach experiment therefore result from spin quantum number $s = \frac{1}{2}$ producing $2(\frac{1}{2})+1 = 2$ orientations of spin angular momentum. The z-component of spin angular momentum then has allowed values: $m_s = s, s - 1, ..., -s$. For spin quantum number $s = \frac{1}{2}$, m_s is thus limited to the two integer values: $+\frac{1}{2}$ or $-\frac{1}{2}$. Particles with the above characteristics are known as *fermions* obeying Fermi–Dirac statistics in particle physics. The three subatomic particles: protons, neutrons, and electrons, demonstrate this behavior, along with other particles including quarks and neutrinos.

We represent the two orientations of m_s as being eigenvalues of two eigenfunctions: $| \alpha \rangle$ and $| \beta \rangle$. These functions are normalized:

$$\langle \alpha | \alpha \rangle = \langle \beta | \beta \rangle = 1 \tag{11.27}$$

and orthogonal:

$$\langle \alpha | \beta \rangle = \langle \beta | \alpha \rangle = 0 \tag{11.28}$$

The spin wavefunctions thus form an orthonormal eigenvector. The volume element of equations (11.27) and (11.28) is constructed via an abstract concept we will refer to as the spin coordinate. It is now necessary to include spin in addition to three spatial dimensions in the description of a stationary state system.

The spin eigenvectors defined in equations (11.28) and (11.29) in fact have no need for a specific analytic form. The same is true for an operator representing the z-component of a spin angular momentum. We simply require it exhibits the properties:

$$\hat{S}_z|\alpha\rangle = +\frac{1}{2}\hbar|\alpha\rangle \tag{11.29}$$

and

$$\hat{S}_z|\beta\rangle = -\frac{1}{2}\hbar|\beta\rangle \tag{11.30}$$

Again, without substantiation, we introduce the operator $\hat{S}^2$, which acts on the spin eigenfunctions in analogous fashion to $\hat{L}^2$:

$$\hat{S}^2|\alpha\rangle = \hat{S}^2|\beta\rangle = s(s + 1)\hbar^2 \tag{11.31}$$

In keeping with the themes of section 8.2, we expect the total spin and spin component operators to obey the commutation relations: $[\hat{S}^2, \hat{S}_z] = 0$, $[\hat{S}^2, \hat{S}_x] = 0$, and $[\hat{S}^2, \hat{S}_y] = 0$, meaning spin wavefunctions can be simultaneous eigenfunctions of any two. However much like results from chapter 7, equation (7.64), it is expected that: $[\hat{S}_z, \hat{S}_y] = -i\hbar\hat{S}_x$, $[\hat{S}_y, \hat{S}_x] = -i\hbar\hat{S}_z$, and $[\hat{S}_x, \hat{S}_z] = -i\hbar\hat{S}_y$. Only one component of spin angular momentum can be simultaneously measured with $\hat{S}^2$.

The spin eigenfunctions are invariably chosen as eigenfunctions of $\hat{S}^2$ and component $\hat{S}_z$, with spin angular momentum presented by a precessing vector model as shown in figure 11.3. The representation is intended to emphasize there cannot be simultaneous knowledge of the $\hat{S}_x$ or $\hat{S}_y$ components, just like the portrayal of spatial angular momentum in figure 8.3(c). According to equation (11.31) intrinsic or spin angular momentum vector has length: $\sqrt{3}/2 \cdot \hbar$. It is projected along the z-axis by amounts: $\pm 1/2 \cdot \hbar$ according to equations (11.29) and (11.30). In orbital energy diagrams of chemistry, the function $| \alpha \rangle$ is commonly designated by the symbol: ↑ and is referred to as the 'spin up' state as indication of its magnetic north direction. The function $| \beta \rangle$ is then identified by: ↓ and is called the 'spin down' state, representing an opposing magnetic north.

The addition of spin has little impact on the discussion of section 11.1. Since neither the kinetic energy or potential energy operators have terms involving electron spin, The commutation relation $[\hat{S}^2, \hat{H}] = 0$ holds. Therefore, functions of table 11.1 can be simultaneous eigenfunctions of energy along with spatial and spin angular momentum. For a one electron atom, the only modification would be to multiply any of the 1s, 2s, ... spatial wavefunctions by either the function $| \alpha \rangle$ or $| \beta \rangle$. When this is done we form what is referred to as a *spin–orbital*. A large collection of hydrogen atoms each have an electron with a 50:50 chance of being either spin up or spin down. Despite its limited consequence for hydrogen, spin will have profound influence on the many-electron case.

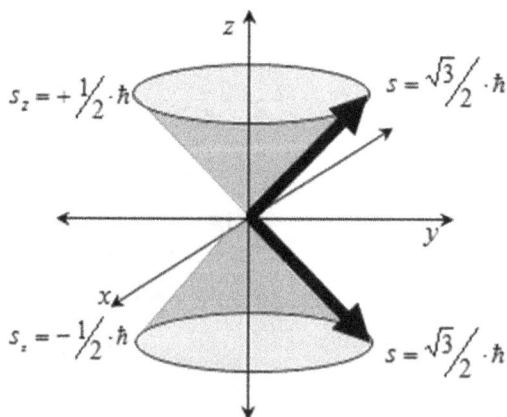

Figure 11.3. The precessing vector model of spin.

11.4 Populating many-electron atoms

Electron occupation of a many-electron atom is conducted according to a few ground rules. The first is the *aufbau prinzip*, which simply states the lowest energy configuration is favored, formed by occupying the available electronic levels of lowest energy. At this point electronic energy is solely determined by principal quantum number n, so it would seem that electrons have a large number of equal energy choices, based on the number of allowed orbitals and their degeneracies. We will return to this point in a moment.

The second occupation rule is the *Pauli exclusion principle*, which asserts no two electrons on the same atom can simultaneously possess the same four quantum number values. Envisioning quantum numbers from the viewpoint of an electron behaving as a particle, we may think of them like spatial coordinates. The Pauli principle thus implies matter occupies its own unique region of space. For example, two individuals attending a football game can enter the stadium through the same gate, go to the same section, and even climb to the same row so long as their seat numbers differ. If instead we view from the perspective of waves, quantum numbers adjust each electron's phase so they do not exhibit interference. Based on the Pauli exclusion principle, we conclude that the first energy level can hold two electrons, as the only available orbital is the non-degenerate 1s which is further allowed a spin up and down electron. On the second level there again is an s but now also a triply degenerate p set with a total number of electrons capped at eight. Continuing this approach, energy level three has maximum capacity: 18 electrons, level four: 32 electrons, etc.

Figure 11.4(a) depicts the distribution of available electronic orbitals based on the one electron hydrogen atom solution of section 11.1. According to equation (11.9) or (11.10), levels increase in energy (in fact, become less negative) as a function of n. On each successive level there is an additional orbital available orbital of $2\ell + 1$ degeneracy. The key feature of the hydrogen atom solution is that all orbitals of a given n value possess the same energy independent of their angular quantum number ℓ, magnetic angular quantum number m_ℓ or magnetic spin quantum number m_s.

Although it is logical to interpret many electron atoms using the solution of hydrogen, as we begin to populate electrons beyond the $n = 1$ energy level we encounter major differences between this model and actual cases. On a given level, the degeneracy between available orbitals with different ℓ values is broken. Reasons behind this can be argued from a mixture of classical and quantum mechanical concepts. Electrons in many-electron atoms experience Coulombic repulsive forces due to one another. According to discussions in section 11.1, electrons on sequential energy levels have the same increasing average distance from the nucleus. As a result, electrons on the nth energy level are *shielded* or *screened* from the full nuclear charge Z by electrons on the $n - 1$ $n - 2$, ... energy levels. Electrons populating levels beyond $n = 1$ therefore experience an *effective nuclear charge*:

$$Z_{\text{eff}} = Z - \sigma \tag{11.32}$$

where σ is the *screening constant* for electrons on a particular energy level. Many sources call this the *shielding constant*, but there is too close an association of that

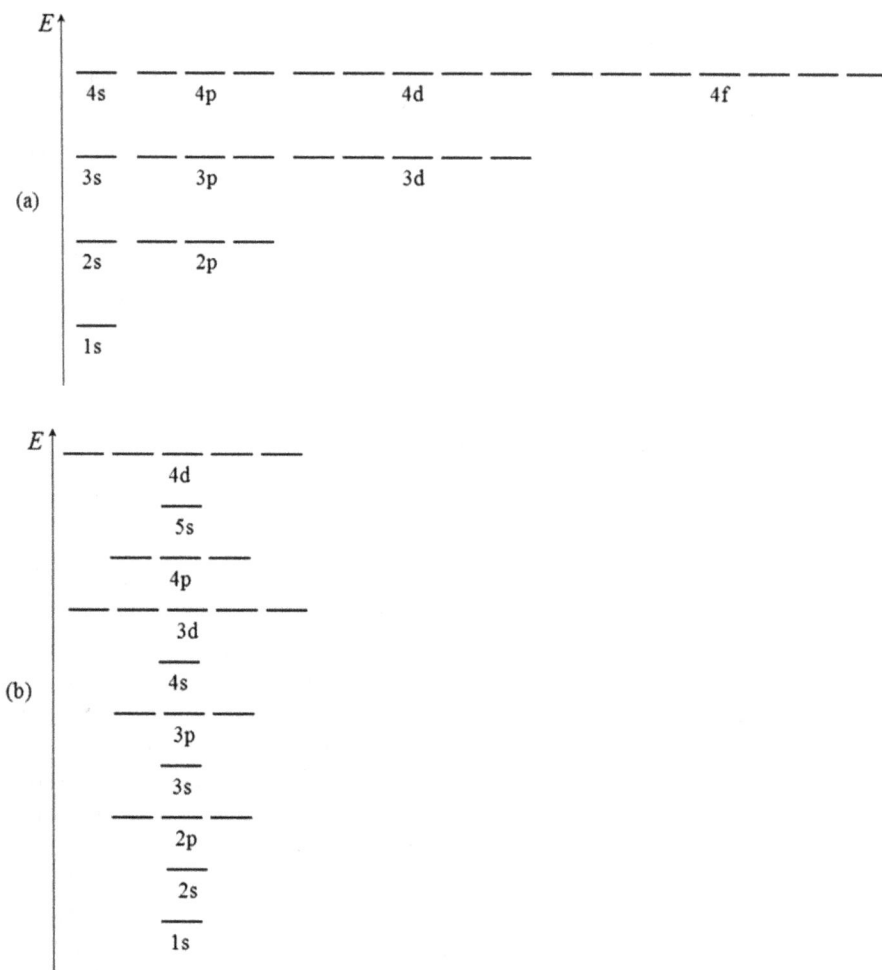

Figure 11.4. (a) Energy diagram for the one-electron atom. (b) Energy diagram for the many-electron atom.

term to the magnetic field effect a nucleus experiences in a magnetic resonance experiment.

Electrons possessing the same n but differing ℓ values have the degeneracy of their energies broken by the presence of other electrons. This can be explained by examining radial wavefunctions in table 11.1. Hydrogenics with the same n but differing ℓ values have different radial distributions. A more revealing indicator is the radial probability density plots for $n = 3$ hydrogenics displayed in figure 11.2(d). Because their probability densities all reach their greatest maximum at the same r value, electrons in 3s, 3d, and 3p orbitals have the same most probable distance from the nucleus. Compared to the 3d, the 3p displays a second maximum of lower probability at a distance closer to the nucleus. The 3p electron has a finite probability of being at this location, allowing a *penetration of the screening* of core electrons, thus lowering its energy. The 3s electron has a total of three maxima, and enhanced penetration compared to either the 3d or 3p. As a result, in contrast to the behavior of a single electron atom, orbitals of the same n but differing ℓ values have the energy hierarchy: $ns < np < nd < nf$. This results in the familiar pattern of electron population displayed in the periodic table of the elements, and demonstrated in figure 11.4(b).

For computational purposes, hydrogenics represented by equation (11.3) and table 11.1 were historically replaced by *Slater Type Orbitals* (STOs). They have the normalized form:

$$|\psi_n\rangle = \frac{[2\xi/a_0]^{n+1/2}}{[(2n)!]^{1/2}} r^{n-1} e^{-\xi r/a_0} \cdot Y_\ell^{m_\ell}(\theta, \phi) \tag{11.33}$$

Equation (11.33) contains quantities by now familiar such as principal quantum number: n, and Bohr radius: a_0. The exponential factor can be empirically parameterized based on the atomic system. It is often done so in the form:

$$\xi = \frac{Z_{\text{eff}}}{n} = \frac{Z - \sigma}{n} \tag{11.34}$$

where σ is the screening parameter from equation (11.32). Unlike the hydrogenics of table 11.1, the radial functions employed by STOs are not orthogonal. They do however form an orthonormal set based on the properties of either their spherical harmonic function: $Y_\ell^{m_\ell}(\theta, \phi)$ or their spin function. Their efficacy comes in applications to many electron atoms through the exponential factor in equation (11.34).

If desired, the radial components of STOs with different n values but the same ℓ can be Schmidt orthogonalized using the procedure of section 7.3. For instance, the $n = 2$ STO can be orthogonalized to the $n = 1$ function by applying the projection:

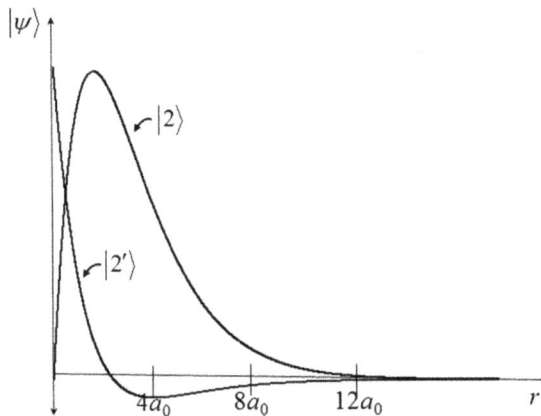

Figure 11.5. Comparison of the original $n = 2$ Slater type orbital: $|2\rangle$ to a Schmidt orthogonalized Slater type orbital: $|2'\rangle = (|2\rangle - |1\rangle\langle 1|2\rangle)/(1 - |\langle 1|2\rangle|^2)^{1/2}$.

$$|2'\rangle = \frac{|2\rangle - |1\rangle\langle 1|2\rangle}{(1 - |\langle 1|2\rangle|^2)^{1/2}} \tag{11.35}$$

It is interesting to contrast amplitude plots of STOs and hydrogenics. Figure 11.5 shows the radial amplitude component of the $n = 2$ STO (labeled: $|2\rangle$). This should be compared to the radial function: $R_{2s}(r)$ depicted in figure 11.1(b). As equation (11.33) shows, STO $|2\rangle$ lacks the radial node of $R_{2s}(r)$ because it lacks a polynomial form (see table 11.1). However, when the STO $|1\rangle$ is projected out, it now displays a striking similarity to the $R_{2s}(r)$ hydrogenic.

11.5 Many-body wavefunctions

Let us now attempt to analyze a many electron atom from the mathematics of the time independent Schrödinger equation. Solving the eigenvalue problem for a Hamiltonian representing a collection of charged quantum mechanical particles introduces the complexity of an increase in spatial coordinates that also now exhibit interdependence. For instance, consider the simplest possible multi-electron atom, helium. Using the coordinate system shown in figure 11.6, we can write the following electronic Hamiltonian:

$$\hat{H} = \hat{T} + \hat{V} = -\frac{\hbar^2}{2m_e}\hat{\nabla}_1^2 - \frac{\hbar^2}{2m_e}\hat{\nabla}_2^2 - \frac{2e^2}{4\pi\varepsilon_0 R_1} - \frac{2e^2}{4\pi\varepsilon_0 R_2} + \frac{e^2}{4\pi\varepsilon_0 r_{12}} \tag{11.36}$$

An immediate complication arises from the last term of equation (11.36), which depends upon the coupled coordinates of both electron 1 and electron 2, while the kinetic energy and nuclear attraction terms for each electron have separable variables. In fact, it is possible to write the general equation for an n-electron atom:

$$\hat{H} = \hat{H}(1) + \hat{H}(2) + \ldots + \hat{H}(n) + \sum_{i<j} \frac{e^2}{4\pi\varepsilon_0 r_{ij}} \tag{11.37}$$

In equation (11.37), $\hat{H}(1)$ contains the kinetic energy and nuclear attraction terms for electron 1 and has no dependence on the coordinates of the remaining $n-1$

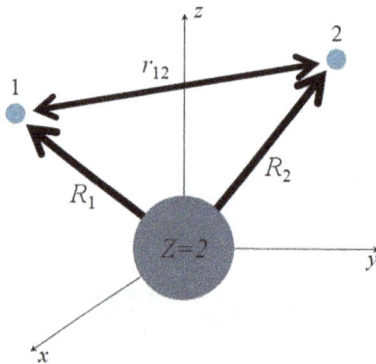

Figure 11.6. Coordinate system for a helium atom.

electrons. However, the summation in equation (11.37) includes $n-1$ terms for the repulsion of electron 1 to all remaining electrons, which makes it impossible to separate dependence of electron 1 from any of the others. Similar arguments render all electronic coordinates interdependent.

A common approach across physics is to seek solutions to simplified model systems, both to explore what information they can provide as well as to establish starting points in a hierarchy of sophistication. As a first approximation to the quantum mechanics of a system of like-charged particles, suppose for the moment that all repulsive electrostatic effects are negligible or can in some way be measured relative to one fixed point in space. The system constitutes a set of *non-interacting* particles, with no *correlation* to their motion. Once an approximate form of this type is found, the correct treatment can always be added back in as a *perturbation*, using techniques described in chapter 12. By eliminating or parameterizing all electron–electron repulsion terms in equation (2.37), the kinetic energy and nuclear attraction terms for each particle are independent functions of individual coordinates. This allows a separation of variables as was done for the 2-dimensional particle in a box in section 6.3.

Let us assume we replace the electrons in our many-body problem by charge-less masses tiny enough so that gravitation effects can be summarily ignored. It is possible the system is under the influence of some type of potential, but let us further assume any effect it has on each particle is separable. The n-particle Hamiltonian is now:

$$\hat{H} = \hat{H}(1) + \hat{H}(2) + \ldots + \hat{H}(n) \tag{11.38}$$

The n-particle many-body wavefunction can thus be expressed as a product of n one-particle wavefunctions. If formed from a basis of orthonormal functions, it is written:

$$|\psi(1, 2, \ldots, n)\rangle = |\psi(1) \cdot \psi(2) \cdot \ldots \cdot \psi(n)\rangle \tag{11.39}$$

Using equations (11.38) and (11.39) with $n = 2$, the eigenvalue problem is:

$$\hat{H}|\psi(1, 2)\rangle = (\hat{H}(1) + \hat{H}(2))|\psi(1) \cdot \psi(2)\rangle$$
$$= |\psi(2) \cdot \hat{H}(1)\psi(1)\rangle + |\psi(1) \cdot \hat{H}(2)\psi(2)\rangle \tag{11.40}$$

The final identity above follows from the fact that functions 1 and 2 have independent coordinates. If $\psi(1)$ and $\psi(2)$ are eigenfunctions to their respective Hamiltonians with eigenvalues $E(1)$ and $E(2)$ then:

$$\hat{H}|\psi(1, 2)\rangle = E(1)|\psi(1) \cdot \psi(2)\rangle + E(2)|\psi(1) \cdot \psi(2)\rangle$$
$$= E(1, 2)|\psi(1, 2)\rangle \tag{11.41}$$

In general, if the Hamiltonian can be written as a sum of Hamiltonians, the system wavefunction is a product function and the total energy is a sum of individual energies:

$$E(1, 2, \ldots, n) = E(1) + E(2) + \cdots + E(n) \tag{11.42}$$

When evaluating an expectation value, the implication of separable electronic coordinates is as follows:

$$\langle o(1, 2, ..., n)\rangle = \langle \psi(1) \cdot \psi(2) \cdot ... \psi(n) | \hat{O}(1) + \hat{O}(2) + ... + \hat{O}_n |$$
$$\psi(1) \cdot \psi(2) \cdot ... \psi(n)\rangle$$
$$= \langle \psi(1) | \hat{O}(1) | \psi(1)\rangle(\langle\psi(2)|\psi(2)\rangle...\langle\psi(n)|\psi(n)\rangle) \qquad (11.43)$$
$$+ \langle \psi(2) | \hat{O}(2) | \psi(2)\rangle(\langle\psi(1)|\psi(1)\rangle...\langle\psi(n)|\psi(n)\rangle) + ...$$
$$= \langle o(1)\rangle + \langle o(2)\rangle + ... + \langle o(n)\rangle$$

Our product wavefunction resulted from a separation of variables by approximating a many-body problem of non-interacting electrons. Although this approach is too drastic for most practical applications, it nonetheless retains widespread acceptance for symbolically representing electron configurations in freshman chemistry. Based on the *aufbau prinzip*, the Pauli exclusion principle, and the screening/penetration effect, the many-electron configuration of, for instance iron, is familiarly written:

$$1s^2 2s^2 2p^6 3s^2 3p^6 4s^2 3d^6 \qquad (11.44)$$

which we now recognize as being a product function of hydrogenic wavefunctions for non-interacting electrons.

Although it is impossible to analytically solve the Schrödinger equation for an atom or molecule containing two or more interacting electrons, numerical models can approximate the exact solution to a high degree of accuracy. The most popular of these is *self-consistent field* (SCF) theory. Also known as a mean field theory, it was first developed and applied to product wavefunctions by Hartree. His approach is a variational method (see chapter 10), finding an upper bound to the exact electronic wavefunction by optimizing a basis of orthonormal spin orbitals that are eigenfunctions of an approximate Hamiltonian of one-electron operators:

$$\hat{H} = \sum_i \hat{H}_i = \sum_i \left(-\frac{\hbar^2}{2m_e}\hat{\nabla}_i^2 - \frac{Ze^2}{4\pi\varepsilon_0 R_i} + \hat{v}_i^{\mathrm{avg}} \right) \qquad (11.45)$$

In equation (11.45), the Hamiltonian for each electron is made separable from every other by replacing the exact electron–electron repulsion by an average repulsion experienced by electron i due to the presence of all others. The average potential is the approximate repulsion an electron experiences due to the most likely location of all others. It is logical to use probability densities to represent electron locations, hence:

$$\hat{v}_i^{\mathrm{avg}} = \frac{e^2}{4\pi\varepsilon_0}\sum_{j\neq i}\int \frac{|\psi_j|^2}{r}\mathrm{d}\tau_j \qquad (11.46)$$

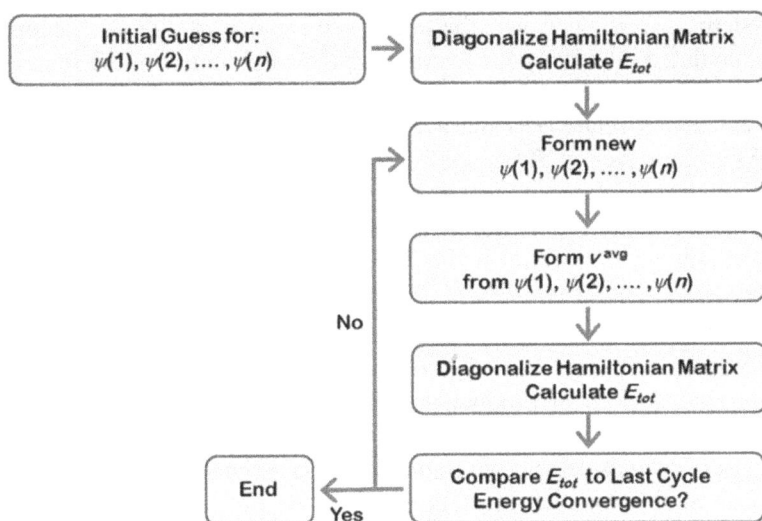

Figure 11.7. The algorithm for a self-consistent field method.

Hence the repulsion an electron experiences is dependent on the manner in which the orbitals describing its neighbors are constructed. The one-electron Hamiltonians are therefore dependent on their own eigenfunctions. The equations are therefore non-linear, and must be solved iteratively.

Figure 11.7 diagrams the SCF algorithm. It begins with an initial guess for the total wavefunction. A product of basis set wavefunctions for non-interacting electrons is one possibility. If this is the case, a matrix with elements constructed from the one-electron Hamiltonians with no average electron repulsion potential. Expansion coefficients for the basis functions are then determined using matrix equations according to the variational recipes of section 10.3:

$$\mathbf{H\,C} \;=\; \mathbf{S\,C\,E} \tag{11.47}$$

The total energy of the initial guess is found by summing eigenvalues of occupied orbitals.

With coefficients from equation (11.47), the average repulsion $\hat{v}_i^{\text{avg}}$ is constructed for each one electron Hamiltonian $\hat{H}_i$, giving a second Hamiltonian matrix to diagonalize. When the ground state energy is determined from this step, it is compared to the value from the previous diagonalization. If their difference is within a desired tolerance, the wavefunction has been optimized and a *self-consistent field* has been reached for the potential. If the energy is not adequately converged, the most recent $\mathbf{C}$ matrix is used to form another $\hat{v}_i^{\text{avg}}$ and $\hat{H}_i$, and the process continues. As an interesting historical note, Hartree devised this procedure prior to the availability of computers. This would seem an impossible task without their number crunching ability, however Hartree's father was a retired accountant, inured to the tedium of mountainous sets of numbers. He played the computer's role, and was invaluable to the successful optimization of the first SCF wavefunctions.

In a system with n electrons, the energy eigenvalue of the nth occupied SCF orbital is calculated relative to the average repulsion potential of the remaining $n-1$ occupied orbitals. In addition, the orbital energy of the first unoccupied, or *virtual* orbital is calculated relative to the average repulsion potential of the n occupied orbitals. As a consequence, *Koopman's Theorem* states that the negative eigenvalue (the sign is changed because it is negative for bound states) of the highest occupied orbital represents the first ionization potential of the system, and the eigenvalue of the lowest energy virtual orbital is the SCF electron affinity.

11.6 Antisymmetry

There yet remains one more complication inherent to a many-electron wavefunction. The non-correlated electron product wavefunctions we have constructed made no mention of any spin coordinate other than indicating the maximum occupation of a degenerate set of orbitals. When spin is introduced to the product wavefunction, there is an additional condition which must be satisfied. As previously mentioned, particles possessing half-integer spin are classified as fermions. A many-body system composed of fermions must obey the *spin statistics theorem* first proposed by Fierz but reformulated by Pauli. This states that fermions must be described by wavefunctions that are *antisymmetric* with respect to the *permutation* or interchange of any two particles:

$$| \psi(1, 2, ..., i, j, ..., n)\rangle = -| \psi(1, 2, ..., j, i, ..., n)\rangle \tag{11.48}$$

Equation (11.48) is equivalent to a 180° change in phase of wave amplitude. Permuting particles does not affect the sign of any expectation value, which depends upon a wavefunction's probability density. This is the modulus square of the wavefunction amplitude. Thus, reversing phase has no effect on this quantity, no matter if it is obtained from a real, imaginary or complex wavefunction.

To satisfy the stipulation of equation (11.48), a many-electron wavefunction must be *antisymmetrized*. For example, consider a system of two non-interacting fermions. The total wavefunction must consist of both spatial and spin components constructed so that equation (11.48) is satisfied. If the particles have opposing spins the total wavefunction is constructed as a product of symmetric spatial and antisymmetric spin wavefunctions:

$$|\psi(1, 2)\rangle = |\psi(1) \cdot \psi(2)\rangle \cdot \frac{1}{\sqrt{2}}(|\alpha(1) \cdot \beta(2)\rangle - |\beta(1) \cdot \alpha(2)\rangle) \tag{11.49}$$

The factor: $1/\sqrt{2}$ normalizes $| \psi(1, 2)\rangle$. If the particles possess the same spin, the spatial component carries antisymmetry:

$$|\psi(1, 2)\rangle = |\alpha(1) \cdot \alpha(2)\rangle \cdot \frac{1}{\sqrt{2}}(|\psi(1) \cdot \psi(2)\rangle - |\psi(1) \cdot \psi(2)\rangle) \tag{11.50}$$

PARALLEL INVESTIGATION: Verify that equations (11.43) and (11.44) exhibit the property: $|\psi(1, 2)\rangle = -|\psi(2, 1)\rangle$

For instance, the wavefunction representing spin state: $\uparrow\downarrow\uparrow$ would be:
$$\underset{1s\ 2s}{}$$

$$|\psi(1, 2, 3)\rangle = \frac{1}{\sqrt{6}}$$

$$\times \left(\begin{array}{l} |\psi_{1s}(1)\alpha(1)\psi_{1s}(2)\beta(2)\psi_{2s}(3)\alpha(3)\rangle + |\psi_{1s}(2)\alpha(2)\psi_{1s}(3)\beta(3)\psi_{2s}(1)\alpha(1)\rangle \\ +|\psi_{1s}(3)\alpha(3)\psi_{1s}(1)\beta(1)\psi_{2s}(2)\alpha(2)\rangle - |\psi_{1s}(3)\alpha(3)\psi_{1s}(2)\beta(2)\psi_{2s}(1)\alpha(1)\rangle \\ -|\psi_{1s}(1)\alpha(1)\psi_{1s}(3)\beta(3)\psi_{2s}(2)\alpha(2) - |\psi_{1s}(2)\alpha(2)\psi_{1s}(1)\beta(1)\psi_{2s}(3)\alpha(3)\rangle\rangle \end{array} \right) \quad (11.51)$$

It is obvious that this procedure becomes unruly in very short order. Once again, linear algebra supplies an answer, providing a matrix representation of the anti-symmetrized wavefunction in the concise form of a *Slater determinant*. For instance, a 2×2 matrix has determinant defined:

$$\det \begin{vmatrix} a & b \\ c & d \end{vmatrix} = ad - bc \quad (11.52)$$

The determinant of any higher order square matrix is found using the reduction formula:

$$\det \begin{vmatrix} a & b & c \\ d & e & f \\ g & h & i \end{vmatrix} = a \cdot \det \begin{vmatrix} e & f \\ h & i \end{vmatrix} - b \cdot \det \begin{vmatrix} d & f \\ g & i \end{vmatrix} + c \cdot \det \begin{vmatrix} d & e \\ g & h \end{vmatrix} \quad (11.53)$$

$$= a(ei - fh) - b(di - fg) + c(dh - eg)$$

Assuming a *closed-shell* (all paired electrons) n electron case, a Slater determinant is then constructed from the recipe:

$$|\psi(1, 2, \ldots n)\rangle = \frac{1}{\sqrt{n!}} \det \begin{vmatrix} \psi_a(1)\alpha(1) & \psi_a(2)\alpha(2) & \psi_a(3)\alpha(3) & \ldots & \psi_a(n)\alpha(n) \\ \psi_a(1)\beta(1) & \psi_a(2)\beta(2) & \psi_a(3)\beta(3) & \ldots & \psi_a(n)\beta(n) \\ \psi_b(1)\alpha(1) & \psi_b(2)\alpha(2) & \psi_b(3)\alpha(3) & \ldots & \psi_b(n)\alpha(n) \\ \vdots & \vdots & \vdots & & \vdots \\ \psi_z(1)\beta(1) & \psi_z(2)\beta(2) & \psi_z(3)\beta(3) & \ldots & \psi_z(n)\beta(n) \end{vmatrix} \quad (11.54)$$

In equation (11.54): $1/\sqrt{n!}$ is a normalizing factor. The same approach may also be applied to an open shell case, as is explored below.

PARALLEL INVESTIGATION: Verify by setting up the appropriate 3×3 determinant that equation (11.45) is obtained.

The properties of linear algebra that determinants exhibit make them ideal for representing fermion wavefunctions. If two columns or rows of a matrix are the same, the matrix has zero determinant:

$$\det \begin{vmatrix} a & a \\ c & c \end{vmatrix} = ac - ca = 0 \quad (11.55)$$

This is equivalent to making the spin and spatial wavefunctions of any two columns of equation (11.54) into functions of the same electron coordinate. Doing so would mean that two electrons simultaneously have the same four quantum numbers. This is a violation of the Pauli exclusion principle. Secondly, consider the determinant of the matrix below:

$$\det \begin{vmatrix} b & a \\ d & c \end{vmatrix} = bc - ad \tag{11.56}$$

Compared to equation (11.52), the matrix in equation (11.56) has interchanged its two columns. The calculations show the two matrices have determinants that are opposite in sign. This is generally true for the determinant of a square matrix of any dimension. In equation (11.54), interchanging two columns is equivalent to permuting two electrons' coordinates, which changes the wavefunction's sign.

Relatively simple results are obtained when expectation values of one-electron operators, such as kinetic energy and nuclear attraction, are evaluated for orthonormal spin wavefunctions expressed in a properly antisymmetrized form. Suppose we construct the determinantal function: $|\Psi\rangle$ from orthonormal basis of functions: $|\psi_i\rangle$, and use it to determine a property from a sum of one electron operators as represented by equation (11.38). The expectation value is:

$$\langle \Psi | \hat{H} | \Psi \rangle = \sum_i \langle \psi_i | \hat{H}(i) | \psi_i \rangle \tag{11.57}$$

This is the same result that was found for the simple product wavefunction in equation (11.43). Now suppose an antisymmetrized product: $|\Psi'\rangle$ is formed using the same orthonormal basis, and only differs from $|\Psi\rangle$ by replacing function: $|\psi_i\rangle$ by: $|\psi_j\rangle$ for the third electron. The expectation value formed between these two determinants is:

$$\langle \Psi | \hat{H} | \Psi' \rangle = \langle \psi_i | \hat{H}(3) | \psi_j \rangle \tag{11.58}$$

The expectation value is now a single matrix element. If two determinants differ by two or more basis functions, any expectation value over one-electron operators is zero. These results are also obtained for simple product wavefunctions which ignore antisymmetry that differ by one, two, or more basis functions.

It is also of interest to examine the properties of a two-electron operator, in particular the electron–electron repulsion terms for a Hamiltonian of interacting particles. Using the same definitions of determinants as was done for a one-electron operator, we find:

$$\left\langle \Psi \left| \frac{e^2}{4\pi\varepsilon_0 r} \right| \Psi \right\rangle = \frac{1}{2} \frac{e^2}{4\pi\varepsilon_0} \sum_i \sum_j \left(\begin{array}{l} \int d\tau_1 \int d\tau_2 \psi_i^*(1)\psi_i(1)\frac{1}{r_{12}}\psi_j^*(2)\psi_j(2) \\ - \int d\tau_1 \int d\tau_2 \psi_i^*(1)\psi_j(1)\frac{1}{r_{12}}\psi_j^*(2)\psi_i(2) \end{array} \right) \tag{11.59}$$

The factor: 1/2 is included on the right-hand side to avoid double counting as the sum runs over all basis functions. The interpretation of the top integral of the right-hand side is straightforward. It is called a *Coulomb integral* and represents electrostatic repulsion of electron i's probability density to that due to electron j. Again, this term would arise in the exact same summation form if the wavefunction were a simple product. However, the bottom integral of equation (11.59) does not arise for the simple product. It is referred to as an *exchange integral*, and lacks a simple classical interpretation. Its sign is opposite that of the Coulomb term, allowing electrons with parallel spins to permute in an antisymmetrized wavefunction. For completeness we also identify results for two determinants differing by a single basis function:

$$\left\langle \Psi \left| \frac{e^2}{4\pi\varepsilon_0 r} \right| \Psi' \right\rangle = \frac{e^2}{4\pi\varepsilon_0} \sum_k \left(\begin{array}{c} \int d\tau_1 \int d\tau_2 \psi_i^*(1)\psi_j(1)\frac{1}{r_{12}}\psi_k^*(2)\psi_k(2) \\ - \int d\tau_1 \int d\tau_2 \psi_j^*(1)\psi_k(1)\frac{1}{r_{12}}\psi_k^*(2)\psi_j(2) \end{array} \right) \qquad (11.60)$$

and two determinants that differ by two orbitals:

$$\left\langle \Psi \left| \frac{e^2}{4\pi\varepsilon_0 r} \right| \Psi'' \right\rangle = \frac{e^2}{4\pi\varepsilon_0} \left(\begin{array}{c} \int d\tau_1 \int d\tau_2 \psi_i^*(1)\psi_j(1)\frac{1}{r_{12}}\psi_k^*(2)\psi_l(2) \\ - \int d\tau_1 \int d\tau_2 \psi_i^*(1)\psi_l(1)\frac{1}{r_{12}}\psi_k^*(2)\psi_j(2) \end{array} \right) \qquad (11.61)$$

Determinants that differ beyond two locations lead to expectation values of zero.

The antisymmetry property for many-electron wavefunctions was implemented by Fock into self-consistent field theory, as an improvement to the product wavefunctions first numerically determined by Hartree. The method is now known as the *Hartree–Fock Self-Consistent Field* (HF SCF) method.

11.7 Angular momentum in many-electron atoms

Consider a collection of k electrons, each possessing an individual spin angular momentum of magnitude: $s = +\frac{1}{2}$. The spin of each can be represented by a vector of magnitude: $|\vec{s}| = \sqrt{3}/2 \cdot \hbar$ in accordance with the discussion of section 11.3. We can then take a vector sum of all spins in the system:

$$\vec{S} = \sum_k \vec{s}_k \qquad (11.62)$$

and determine its magnitude. In the case of any completely occupied orbital or set of orbitals, the magnitude of the total spin vector must be:

$$|\vec{S}| = 0 \qquad (11.63)$$

This is explained by referring to figure 11.3. Take a completely occupied set of p orbitals for instance, which has three spin up α electrons with $\vec{s}$ having positive projection along the z-axis and three spin down β electrons with $\vec{s}$ having negative projection along the z-axis. The vector sum of three parallel combined with three antiparallel components is zero. We conclude that equation (11.63) holds for any closed shell or completely occupied non-degenerate or degenerate set of orbitals.

For orbital occupations that are not closed shell, it is convenient to define a total z-projection of angular momentum:

$$M_S = \sum_k m_{s_k} \tag{11.64}$$

This in fact is completely general, because equation (11.64) for any closed shell case is zero. If we now consider the set of p orbitals occupied by two electrons of the same spin, which must enter into two separate p orbitals of the degenerate set to satisfy the Pauli exclusion principle. In this case the magnitude of the total spin vector is: $S = 1$. If the two electrons were of the α type, we then have: $M_S = 1$, but we have not specified if the two electrons were both spin up or spin down. There should be no preference for one over the other in the absence of some external field, so our model should incorporate either case as an equal likelihood. Drawing on conclusions from the one-electron case, we define allowed values of the z-projection of total spin angular momentum to have allowed values:

$$M_S = S, S - 1, ..., -S \tag{11.65}$$

with a total spin vector of magnitude: $|\vec{S}| = \sqrt{S(S + 1)} \cdot \hbar$. Spin states belonging to the same S value are degenerate, with the number of equal energy states known as the *multiplicity*

$$2S + 1 \tag{11.66}$$

The case of two electrons occupying a degenerate set of spatial orbitals has multiplicity: $2 \cdot 1 + 1 = 3$, and is known as a *triplet* spin state. This would be represented by the three possible orientations of combined z spin components: $M_S = +1, 0, -1$. A wavefunction for this state requires an antisymmetric spatial wavefunction, of the form we discussed in equation (11.50). This is multiplied by one of three symmetric spin functions:

$$\alpha(1)\alpha(2)$$
$$\frac{1}{\sqrt{2}}(\alpha(1)\beta(2) + \beta(1)\alpha(2)) \tag{11.67}$$
$$\beta(1)\beta(2)$$

In the normal case there are several different orbitals occupied. The multiplicity of a many-electron atomic system is a product of the multiplicities of all its individual occupied orbital types. Fortunately, all closed shells have $2(0) + 1 = 1$ or *singlet* multiplicity. Although partially filled orbitals can also form singlet spin states, a wide variety of other spin states are common to atomic systems. The *doublet* is any configuration with a lone unpaired electron, including the hydrogen atom. A quartet p state arises from the equivalence of four different spin orientations, which can be symbolically represented:

$$\uparrow\uparrow\uparrow \quad \uparrow\uparrow\downarrow \quad \uparrow\downarrow\downarrow\downarrow\downarrow \tag{11.68}$$

A set of degenerate orbitals that are partially occupied can lead to a variety of spin states, dependent upon the electrons having opposing m_s values in the same orbital, or entering into separate ones with arbitrary m_s. From the many possibilities this creates, the lowest energy spin configuration is given by *Hund's rule*, which states that maximum multiplicity is favored. For example, five electrons placed into a set of d-orbitals are capable of generating a sextet, quartet, or doublet spin state with five, three and one unpaired electron respectively. Hund's rule states that the sextet is lowest in energy, followed by the quartet, and the doublet is highest in energy.

PARALLEL INVESTIGATION: Verify that four electrons placed into a set of d-orbitals generate a possible pentet, triplet, and singlet spin state.

Spatial angular momentum for a many-electron atom may be treated in the same manner. A magnitude for the total orbital angular momentum vector is defined to be: L, along with a combined z-component:

$$M_L = \sum_i m_\ell(i) \tag{11.69}$$

The combined z-component is related to the total spatial angular momentum via: $M_L = L, L-1, ..., -L$. In addition, we introduce a letter symbolic of total orbital angular momentum in accordance with that which was done for one-electron orbital angular momentum in equation (11.16), and simply replaces the lower-case designations by upper case letters:

L	0	1	2	3	4	5	6	7	8	9
letter	S	P	D	F	G	H	I	K	L	M

$$(11.70)$$

As was found in the spin case, it is true that any degenerate set of atomic orbtals that are completely occupied has: $L = 0$ and is designated as an S orbital angular momentum state. It is a partially-occupied degenerate orbital which leads to a variety of spatial angular momentum states. For instance, placing two electrons in a set of d orbitals generates the possibility of a G, F, D, P, and S orbital angular momentum states.

PARALLEL INVESTIGATION: Verify that putting three electrons in a set of d orbitals generates the possibility of an H, G, F, D, and P state

---

In either a single-electron or multi-electron case, electron spatial and spin angular momentum can interact constructively or destructively, creating slightly altered energies for atomic systems. When coupling the two forms, there is dependence on the spin–orbit coupling that the system exhibits, which dramatically complicates the issue. Here we will adapt the *Russell–Saunders coupling* scheme, in which spin orbit effects are assumed to be small enough to be ignored. This is generally the case for the first few rows of the periodic table. We will then introduce: J as a means to measure the combination of spin and spatial angular momentum. Permitted values of J are given by the *Clebsch–Gordon series*:

$$J = L + S, L + S - 1, ..., |L - S| \tag{11.71}$$

A singlet spin state of orbital angular momentum corresponding to S has $J = 0$, the obvious case of no angular momentum to couple. In fact, according to equation (11.71) any S type orbital state or singlet spin state posseses one value of coupled angular momentum. A triplet P state on the other hand possesses J values, 2, 1, and 0.

PARALLEL INVESTIGATION: Verify that a quartet D state has allowed values of coupled angular momentum: $J = 7/2, 5/2, 3/2, 1/2$.

Information from spatial, spin, and coupled angular momentum culminates in a *term symbol*, which uses a letter designation representing the total orbital angular momentum, a left superscript describing spin multiplicity, and a right subscript showing which particular J state is being referenced. For example, the state of no net spin or orbital angular momentum has the term symbol: $^1S_0$, while there are four possible term symbols for the $L = 3$, $S = 2$ case each representing the various ways in which the spin and orbital angular momentum interact. These are: $^5F_5$, $^5F_4$, $^5F_3$, and $^5F_2$.

PARALLEL INVESTIGATION: Verify the possible term symbols for the triplet D state are: $^3D_3$, $^3D_2$, and $^3D_1$.

The transitions which occur between electronic states in atomic spectroscopy are described with term symbols. Using expectation values, a set of *selection rules* have been established to indicate which states are allowed to couple upon absorption of a photon. These are:

$$\Delta S = 0 \quad \Delta L = 0, \pm 1 \quad \Delta \ell = \pm 1 \quad \Delta J = 0, \pm 1,$$
$$\text{but } J = 0 \rightarrow J = 0 \text{ forbidden} \tag{11.72}$$

The rule regarding spin reflects the fact that photons are spinless. The rules governing orbital angular momentum ensure that absorbing a photon promotes the electron from a degenerate set. As nuclei increase in Z value, there is a breakdown of the Russell–Saunders scheme and selection rules in equation (11.72) fail. In these cases, effects such as phosphorescence occur, which is a coupling between singlet and triplet spin states.

IOP Concise Physics

What's the Matter with Waves?
An introduction to techniques and applications of quantum mechanics
William Parkinson

Chapter 12

Perturbation theory

12.1 Rayleigh Schrödinger perturbation theory

In the laboratory, atomic and molecular properties are extracted by applying external probes, the most common of these being static electric or magnetic fields or electromagnetic radiation. In theory, response of the system to these disturbances can be calculated by collapsing the wavefunction to a particular eigenstate. To induce this event, a quantum mechanical technique will be used which in spirit is analogous to experiment. We will measure the response of the wavefunction to an additional term, or *perturbation* to its Hamiltonian. For simplicity, we will begin with a static, non-oscillatory external stimulus, and therefore present time-independent perturbation theory.

The process begins by modifying the system's original Hamiltonian $\hat{H}_0$ to introduce some new effect:

$$\hat{H} = \hat{H}_0 + \lambda \hat{V} \tag{12.1}$$

In equation (12.1), $\hat{V}$ is now an operator representing an external perturbation beyond the normal potential that the system experiences. This is a rather unfortunate convention that is common to the development. Note this component is a potential in *addition* to say, a conservative force field as for the harmonic oscillator or the electrostatics of Coulombic attraction for the hydrogen electron, which are effects implicitly included in $\hat{H}_0$. The value λ is an *ordering parameter*, which has no physical meaning but plays a bookkeeping role as will be seen. Note that $\hat{H}_0$ has no ordering parameter dependence so, as the subscript indicates, is known as the *zeroth-order Hamiltonian*.

It is assumed there exists a complete orthonormal set of solutions to $\hat{H}_0$:

$$\hat{H}_0 \, |n\rangle = E_n^{(0)} \, |n\rangle \tag{12.2}$$

with a lowest energy $(E_0^{(0)})$ eigenfunction: $|0\rangle$. The set comprises the *unperturbed* or *zeroth-order* solutions, as categorized by the energy superscript. The system relaxes

doi:10.1088/978-1-6817-4577-0ch12

to the additional Hamiltonian term, producing a ground state *response function*: $|\Phi_0\rangle$, which can also be expanded in a power series in the ordering parameter:

$$|\Phi_0\rangle = |0\rangle + \lambda \left|\Phi_0^{(1)}\right\rangle + \lambda^2 \left|\Phi_0^{(2)}\right\rangle + \ldots \tag{12.3}$$

The ground state response function is specifically constructed to be *intermediately normalized* to the zeroth-order unperturbed ground state:

$$\langle 0|\Phi_0\rangle = 1 \tag{12.4}$$

Multiplying equation (12.3) by ket: $\langle 0 |$ and integrating, we use the normalization of the zeroth-order ground state and equation (12.4) to conclude that:

$$\left\langle 0\middle|\Phi_0^{(1)}\right\rangle = \left\langle 0\middle|\Phi_0^{(2)}\right\rangle = \ldots = 0 \tag{12.5}$$

The perturbed ground state energy is similarly expanded:

$$E_0 = E_0^{(0)} + \lambda E_0^{(1)} + \lambda^2 E_0^{(2)} + \ldots \tag{12.6}$$

All that now remains is to combine equations (12.1), (12.3), and (12.6) in an eigenvalue equation: $\hat{H}|\Phi_0\rangle = E_0|\Phi_0\rangle$:

$$\begin{aligned}
(\hat{H}_0 + \lambda V)&\left[|0\rangle + \lambda\left|\Phi_0^{(1)}\right\rangle + \lambda 2 \left|\Phi_0^{(2)}\right\rangle + \ldots\right] \\
&= \left(E_0^{(0)} + \lambda E_0^{(1)} + \lambda^2 E_0^{(2)} + \ldots\right)\left[|0\rangle + \lambda\left|\Phi_0^{(1)}\right\rangle + \lambda^2\left|\Phi_0^{(2)}\right\rangle + \ldots\right]
\end{aligned} \tag{12.7}$$

equation (12.7) is left multiplied by ket function $\langle 0 |$ and integrated. Using the consequences of intermediate normalization from equation (12.5), we gather surviving terms order-by-order in the parameter λ. The zeroth-order, or λ-independent equation is:

$$E_0^{(0)} = \langle 0 |\hat{H}_0| 0\rangle \tag{12.8}$$

which is the unperturbed ground state energy. Only one non-zero term of ordering parameter: λ^1 is left on each side of equation (12.7), which gives the energy of first-order perturbation theory:

$$E_0^{(1)} = \langle 0 |\hat{V}| 0\rangle \tag{12.9}$$

Equation (12.9) shows this energy to simply be the expectation value of the unperturbed ground state wavefunction evaluated over the perturbing Hamiltonian.

The process of determining order-by-order expressions for the perturbed energy can be continued to any λ^k desired. For instance, left multiplying equation (12.7) by ket: $\langle 0 |$ and imposing equation (12.5) gives the following expressions for the second- and third-order energies by gathering terms in λ^2 and λ^3 respectively:

$$\begin{aligned}
E_0^{(2)} &= \left\langle 0\middle|\hat{V}\middle|\Phi_0^{(1)}\right\rangle \\
E_0^{(3)} &= \left\langle 0\middle|\hat{V}\middle|\Phi_0^{(2)}\right\rangle
\end{aligned} \tag{12.10}$$

To determine energy beyond order 1 requires a representation of the first- and higher-order ground state response wavefunctions. These are expanded from the complementary elements of the zeroth-order eigenvector, for instance the first-order response function is:

$$\left|\Phi_0^{(1)}\right\rangle = \sum_{n\neq 0} c_n^{(1)} \mid n\rangle \tag{12.11}$$

with similar equations for any higher-order response function. The basis $\mid n\rangle$ constitute an orthonormal set, so the expansion coefficients $c_n^{(1)}$ are related to the response function $\mid \Phi_0^{(1)}\rangle$ in the following fashion:

$$\left\langle n\middle|\Phi_0^{(1)}\right\rangle = \sum_{k\neq 0} c_k^{(1)}\langle n|k\rangle = \sum_{k\neq 0} \delta_{kn}c_k = c_n^{(1)} \tag{12.12}$$

Equation (12.12) uses the Kronecker delta defined in section 7.2. Equation (12.12) allows us to write:

$$\left|\Phi_0^{(1)}\right\rangle = \sum_{n\neq 0} |n\rangle\left\langle n\middle|\Phi_0^{(1)}\right\rangle \tag{12.13}$$

Using the language of section 7.3, the above shows that expansion coefficients of the response function are obtained by the *projection operator*: $\mid n\rangle\langle n\mid$.

Expansion coefficients at arbitrary order k can be determined by again gathering all terms of order: λ^k from equation (12.7). For instance for the first-order response function:

$$\hat{H}_0\left|\Phi_0^{(1)}\right\rangle + \hat{V}|0\rangle = E_0^{(0)}\left|\Phi_0^{(1)}\right\rangle + E_0^{(1)}\mid 0\rangle \tag{12.14}$$

This time however instead of multiplying by the zeroth-order ground state ket, we left multiply equation (12.14) by an arbitrary member of the orthogonal complement: $\langle n\mid$ and integrate giving:

$$E_n^{(0)}\left\langle n\middle|\Phi_0^{(1)}\right\rangle + \langle n|\hat{V}|0\rangle = E_0^{(0)}\langle n|\Phi_0^{(1)}\rangle \tag{12.15}$$

equation (12.15) is rearranged and inserted into equation (12.13):

$$\left|\Phi_0^{(1)}\right\rangle = \sum_{n\neq 0} \frac{|n\rangle\langle n|\hat{V}|0\rangle}{E_0^{(0)} - E_n^{(0)}} \tag{12.16}$$

Comparing equations (12.16) and (12.13) it is apparent that the first-order expansion coefficients are:

$$c_n^{(1)} = \frac{\langle n|\hat{V}|0\rangle}{E_0^{(0)} - E_n^{(0)}} \tag{12.17}$$

When equation (12.16) is inserted in the $E_0^{(2)}$ expression (equation (12.10)), we obtain:

$$E_0^{(2)} = \sum_{n\neq 0} \frac{\langle 0|\hat{V}|n\rangle\langle n|\hat{V}|0\rangle}{E_0^{(0)} - E_n^{(0)}} = \frac{|\langle 0|\hat{V}|n\rangle|^2}{E_0^{(0)} - E_n^{(0)}} \tag{12.18}$$

Equation (12.18) is referred to as a *sum over states* expression.

PARALLEL INVESTIGATION: Verify from equation (12.7) that the response equation of order $k = 2$ is: $\widehat{H_0}|\Phi_0^{(2)}\rangle + \hat{V}|\Phi_0^{(1)}\rangle = E_0^{(0)}|\Phi_0^{(2)}\rangle + E_0^{(1)}|\Phi_0^{(1)}\rangle + E_0^{(2)}|0\rangle$. Next expand the second-order response function in the form of equation (12.13), and left multiply the above by: $\langle n|$ to obtain an expression for the second-order expansion coefficients. Finally, insert these into equation (12.10) to obtain the third-order Rayleigh–Schrödinger perturbation energy:

$$E_0^{(3)} = \sum_{m,n\neq 0} \frac{\langle 0|\hat{V}|n\rangle\langle n|\hat{V}|m\rangle\langle m|\hat{V}|0\rangle}{\left(E_0^{(0)} - E_n^{(0)}\right)\left(E_0^{(0)} - E_m^{(0)}\right)} - E_0^{(1)}\sum_{n\neq 0} \frac{|\langle 0|\hat{V}|n\rangle|^2}{\left(E_0^{(0)} - E_n^{(0)}\right)^2} \quad (12.19)$$

12.2 Applications of perturbation theory

In chapter 10 there are two examples of applying the variational principle on particle in a box wavefunctions to obtain approximate solutions of the harmonic oscillator problem. We will here look at these as an elementary example of perturbation theory. The variational examples separately used sine and cosine trial wavefunctions. For each, the zeroth-order Hamiltonian is the standard particle in a box Hamiltonian containing only a kinetic energy term of the form:

$$\hat{H}_0 = -\frac{\hbar^2}{2\mu}\frac{d^2}{dx^2} \quad (12.20)$$

which for the sine function over the limits: $0 \leqslant x \leqslant a_0$ has a zeroth-order ground state ($n = 1$) energy in atomic units of:

$$E_0 = \langle 0|\hat{H}_0|0\rangle = \frac{h^2 n^2}{8\mu a_0^2} = \frac{\pi^2}{2} = 4.93 \text{ au} \quad (12.21)$$

The additional potential is treated as a perturbation, which in the instance of the sine solutions was:

$$\hat{V} = 2k(x - a_0/2)^2 \quad (12.22)$$

with a force constant: $k = 200$ au. The first-order perturbed energy is given by equation (12.9), which expressed in atomic units is:

$$E_0^{(1)} = \langle 0|\hat{V}|0\rangle = 2 \cdot 2 \cdot 200 \cdot \int_0^1 (x - 1/2)^2 \sin^2(\pi x)dx = 13.07 \text{ au} \quad (12.23)$$

The harmonic oscillator perturbed energy through first-order for the sine function is therefore: $4.93 + 13.07 = 18.0$ au.

The cosine function over the limits: $-a_0 \leqslant x \leqslant a_0$ has a ground state ($n = 1$) zeroth-order energy of:

$$E_0 = \langle 0|\hat{H}_0|0\rangle = \frac{h^2 n^2}{32\mu a_0^2} = \frac{\pi^2}{8} = 1.23 \text{ au} \tag{12.24}$$

The potential in this particular case has the form of the standard harmonic oscillator:

$$V = \frac{1}{2}kx^2 \tag{12.25}$$

Perturbation theory gives a first-order energy in atomic units of:

$$E_0^{(1)} = \langle 0|\hat{V}|0\rangle = \frac{k}{2} \cdot \int_{-1}^{1} x^2\cos^2\left(\frac{\pi x}{2}\right)dx = 13.07 \text{ au} \tag{12.26}$$

Notice this is the same first-order energy obtained for sine function. The harmonic oscillator perturbed energy through first-order for the cosine function is therefore: 1.23 + 13.07 = 14.3 au.

The perturbed energy in either form poorly reproduces the exact harmonic oscillator energy of: $\sqrt{k}/2 = 7.07$ au. A variational treatment of the problem, particularly for the cosine function, is a much better approach in this instance. There is also a sound mathematical argument for this discrepancy. A perturbation series, like a power series, converges rapidly under the conditions that the perturbation is small compared to the zeroth-order expansion point, or that:

$$\langle \hat{H}_0 \rangle \gg \langle \hat{V} \rangle \tag{12.27}$$

A force constant of 200 au was purposely selected in chapter 10 to create a sufficiently deep potential well so that the first four sine or cosine solutions were subject to its influence.

As a second application, consider a perturbative expansion of the dipole operator:

$$\hat{V} = e \cdot \hat{r} \tag{12.28}$$

where e is the unit of elementary charge and $\hat{r}$ is the generalized position operator. Consider this perturbation used to determine the electric field response of hydrogen. The atom is exposed to static electric field components: $\vec{E}_i$, where the subscript i refers to orientation of the field along one of three possible Cartesian directions x, y, or z. This induces a dipole moment $\vec{\mu}_i$ in the system which, provided the field strength is nominal, can be expanded in a power series:

$$\vec{\mu}_i = \vec{\mu}_{0i} + \sum_j \overline{\alpha}_{ij} \cdot \vec{E}_j + \frac{1}{2!}\sum_{j,k} \overline{\beta}_{ijk} \cdot \vec{E}_j \cdot \vec{E}_k + \cdots \tag{12.29}$$

In equation (12.29), $\vec{\mu}_{0i}$ is the field independent, or permanent, dipole moment of the system along Cartesian direction i. The quantities: $\overline{\alpha}, \overline{\beta}, \ldots$ are the polarizability, first hyperpolarizability, etc. These are second-, third-, ... rank tensors characterizing the slope and curvature of the induced dipole moment. For atoms, or any centrosymmetric molecular system for that matter, all odd-order terms in the expansion vanish.

The energy of the system is determined from:

$$\mathcal{E} = -\int \vec{\mu} \cdot d\vec{E} \qquad (12.30)$$

which upon inserting equation (12.29) gives the energy of response to the static electric field:

$$\mathcal{E} = \mathcal{E}_0 - \vec{\mu}_0 \cdot \vec{E} - \frac{1}{2}\overline{\alpha} \cdot \vec{E} \cdot \vec{E} - \frac{1}{6}\overline{\beta} \cdot \vec{E} \cdot \vec{E} \cdot \vec{E} - \dots \qquad (12.31)$$

In equation (12.31), there are implied tensor products for the second and higher order terms.

What we now do is use the electric field as the ordering parameter in equation (12.31), and equate energy terms from perturbation theory at each order. A zero-field integration limit produces the system energy in the absence of electric field, which comparing to equation (12.8) is simply the ground state expectation value:

$$\mathcal{E}_0 = \langle 0|\hat{H}_0|0\rangle \qquad (12.32)$$

Referring to section 11.1, $\hat{H}_0$ can be found in equation (11.6). The zeroth-order energy in atomic units is given by equation (11.10) and has the value: $\mathcal{E}_0 = -0.500$ au.

The permanent dipole moment for non-centrosymmetric systems is matched to the term of equation (12.31) which is first-order in $\vec{E}$, so is therefore calculated by inserting equation (12.28) into equation (12.9). Below this is expressed in atomic units (elementary charge $e = 1$):

$$\mu_{0i} = \langle 0|\hat{r}_i|0\rangle \qquad (12.33)$$

All components of this quantity are zero for the hydrogen atom, as is demonstrated by using $r\cos\theta$ for the spherical polar representation of the z-component along with the $Z = 1$ 1s hydrogenic expression from table 12.1. Applying the spherical polar volume element, we then evaluate the following integral (shown in atomic units):

$$\langle 0|\hat{z}|0\rangle = \left(\frac{1}{\pi}\right)\int_0^{2\pi} d\varphi \int_0^{\pi} \sin(\theta)\cos(\theta)d\theta \int_0^{\infty} r^3 e^{-2r}dr = 0 \qquad (12.34)$$

Comparing equations (12.31) and (12.18), components of the polarizability tensor derive from the second-order perturbation expression when inserting the perturbing operator in equation (12.28) (in atomic units):

$$\alpha_{ij} = -2 \cdot \sum_{n \neq 0} \frac{\langle 0|\hat{r}_i|n\rangle\langle n|\hat{r}_j|0\rangle}{E_0^{(0)} - E_n^{(0)}} \qquad (12.35)$$

The polarizability is of particular interest where the perturbing field is of such small magnitude that higher-order hyperpolarizability field effects are negligible. The polarizability then represents the distortion of electron density surrounding an atom or molecule. For a gas phase sample of randomly-oriented rapidly tumbling

Table 12.1. Normalized real forms for hydrogenics from $n = 1$ to $n = 3$.

$|\psi_{1s}\rangle$	$\dfrac{1}{\sqrt{\pi}}\left(\dfrac{Z}{a_0}\right)^{3/2} \cdot e^{-Zr/a_0}$
$|\psi_{2s}\rangle$	$\dfrac{1}{4\sqrt{2\pi}}\left(\dfrac{Z}{a_0}\right)^{3/2}\left(1 - \dfrac{Zr}{2a_0}\right) \cdot e^{-Zr/2a_0}$
$|\psi_{2p_x}\rangle$	$\dfrac{1}{4\sqrt{2\pi}}\left(\dfrac{Z}{a_0}\right)^{5/2} \cdot re^{-Zr/2a_0}\sin(\theta) \cdot \cos(\varphi)$
$|\psi_{2p_y}\rangle$	$\dfrac{1}{4\sqrt{2\pi}}\left(\dfrac{Z}{a_0}\right)^{5/2} \cdot re^{-Zr/2a_0}\sin(\theta) \cdot \sin(\varphi)$
$|\psi_{2p_z}\rangle$	$\dfrac{1}{4\sqrt{2\pi}}\left(\dfrac{Z}{a_0}\right)^{5/2} \cdot re^{-Zr/2a_0}\cos(\theta)$
$|\psi_{3s}\rangle$	$\dfrac{1}{81\sqrt{3\pi}}\left(\dfrac{Z}{a_0}\right)^{3/2}\left(27 - 18\dfrac{Zr}{a_0} + 2\dfrac{Z^2 r^2}{a_0^2}\right) \cdot e^{-Zr/3a_0}$
$|\psi_{3p_x}\rangle$	$\dfrac{\sqrt{2}}{81\sqrt{\pi}}\left(\dfrac{Z}{a_0}\right)^{5/2}\left(6 - \dfrac{Zr}{a_0}\right) \cdot re^{-Zr/3a_0}\sin(\theta) \cdot \cos(\varphi)$
$|\psi_{3p_y}\rangle$	$\dfrac{\sqrt{2}}{81\sqrt{\pi}}\left(\dfrac{Z}{a_0}\right)^{5/2}\left(6 - \dfrac{Zr}{a_0}\right) \cdot re^{-Zr/3a_0}\sin(\theta) \cdot \sin(\varphi)$
$|\psi_{3p_z}\rangle$	$\dfrac{\sqrt{2}}{81\sqrt{\pi}}\left(\dfrac{Z}{a_0}\right)^{5/2}\left(6 - \dfrac{Zr}{a_0}\right) \cdot re^{-Zr/3a_0}\cos(\theta)$
$|\psi_{3d_{z^2}}\rangle$	$\dfrac{1}{81\sqrt{6\pi}}\left(\dfrac{Z}{a_0}\right)^{7/2} \cdot r^2 e^{-Zr/3a_0}(3\cos^2(\theta) - 1)$
$|\psi_{3d_{xz}}\rangle$	$\dfrac{\sqrt{2}}{81\sqrt{\pi}}\left(\dfrac{Z}{a_0}\right)^{7/2} \cdot r^2 e^{-Zr/3a_0}\sin(\theta) \cdot \cos(\theta) \cdot \cos(\varphi)$
$|\psi_{3d_{yz}}\rangle$	$\dfrac{\sqrt{2}}{81\sqrt{\pi}}\left(\dfrac{Z}{a_0}\right)^{7/2} \cdot r^2 e^{-Zr/3a_0}\sin(\theta) \cdot \cos(\theta) \cdot \sin(\varphi)$
$|\psi_{3d_{x^2-y^2}}\rangle$	$\dfrac{1}{81\sqrt{2\pi}}\left(\dfrac{Z}{a_0}\right)^{7/2} \cdot r^2 e^{-Zr/3a_0}\sin^2(\theta) \cdot \cos(2\varphi)$
$|\psi_{3d_{xy}}\rangle$	$\dfrac{1}{81\sqrt{2\pi}}\left(\dfrac{Z}{a_0}\right)^{7/2} \cdot r^2 e^{-Zr/3a_0}\sin^2(\theta) \cdot \sin(2\varphi)$

[1]$a_0 = \dfrac{(4\pi\varepsilon_0)\hbar^2}{m_e e^2} = 0.529\text{Å}$

molecules the quantity of interest is the isotropic polarizability determined from the trace average of the polarizability tensor. Spherical symmetry allows this quantity to be evaluated as:

$$\alpha = \frac{1}{3}(\alpha_{xx} + \alpha_{yy} + \alpha_{zz}) = \alpha_{zz} \qquad (12.36)$$

Let us evaluate equation (12.35) using: $|0\rangle = |\psi_{1s}\rangle$. For the states $|n\rangle$ we choose the $n = 2$ hydrogenics in their real form as presented in table 12.1. The energy denominators of equation (12.35) are calculated in atomic units using chapter 11, equation (11.10):

$$E_0^{(0)} - E_n^{(0)} = 0.500 - 0.125 = 0.375 \text{ au} \qquad (12.37)$$

The following results should be verified using wavefunctions from table 12.1 and the Cartesian to spherical polar transformation: $z = r\cos(\theta)$:

$$\langle \psi_{1s} | \hat{z} | \psi_{2s} \rangle = \langle \psi_{1s} | \hat{z} | \psi_{2p_x} \rangle = \langle \psi_{1s} | \hat{z} | \psi_{2p_y} \rangle = 0 \qquad (12.38)$$

What remains is the integral:

$$\langle \psi_{1s} | \hat{z} | \psi_{2p_z} \rangle = \frac{1}{\sqrt{\pi}} \frac{1}{4\sqrt{2\pi}} \int_0^{2\pi} d\varphi \int_0^{\pi} \sin(\theta) \cdot \cos^2(\theta) d\theta \int_0^{\infty} r^4 e^{-3r/2} dr \qquad (12.39)$$

$$= 0.7449 \text{ au}$$

PARALLEL INVESTIGATION: Verify that the results: $\langle \psi_{1s} | \hat{z} | \psi_{2p_x} \rangle = \langle \psi_{1s} | \hat{z} | \psi_{2p_y} \rangle = 0.7449$ au using wavefunctions from table 12.1 in an integral similar to equation (12.39) using the operators: $\hat{x} = r \sin(\theta)\cos(\varphi)$ and $\hat{y} = r \sin(\theta)\sin(\varphi)$.

The sum over states expression therefore becomes a single term:

$$\alpha = -2 \cdot \frac{0.7449^2}{-0.375} = -2.959 \text{ au} = 0.435 \text{Å}^3 \qquad (12.40)$$

For a discussion of unit conversion, see appendix A. Notice that polarizability has volume dimensions. This corroborates with earlier discussion of α representing the degree of electron density distortion experienced by an atom or molecule due to an external field. This effect can of course arise by explicitly placing the system in such a laboratory environment, but can also be induced due to the proximity of neighboring influences in a sample as a London dispersion force. The experimental polarizability of the hydrogen atom is 0.666 Å^3, which shows the perturbation theory result is of the correct order of magnitude, but is in error by 34.6%.

PARALLEL INVESTIGATION: Verify using the functions of table 12.1 that including the $n = 3$ s, p, and d hydrogenics in the orthogonal complement $|n\rangle$ leads to only one additional non-zero contributing term: $\langle \psi_{1s} | \hat{z} | \psi_{3p_z} \rangle = 0.2983$ au, and when this is included in the sum over states, leads to a polarizability of: $\alpha = 3.359$ au = 0.494 Å^3, which is in error by 25.8%.

The first hyperpolarizability should be zero by symmetry arguments, which can be corroborated by examining the *zzz* component. This again simplifies greatly due to a large number of vanishing terms. Based on the results obtained during polarizability calculation, the only possible contribution can come from:

$$\beta_{zzz} = -6 \cdot$$

$$\left[\frac{\langle \psi_{1s}|\hat{z}|\psi_{2p_z}\rangle\langle\psi_{2p_z}|\hat{z}|\psi_{2p_z}\rangle\langle\psi_{2p_z}|\hat{z}|\psi_{1s}\rangle}{(E_1 - E_2)^2} - \langle\psi_{1s}|\hat{z}|\psi_{1s}\rangle \cdot \frac{|\langle\psi_{1s}|\hat{z}|\psi_{2p_z}\rangle|^2}{(E_1 - E_2)^2} \right] \qquad (12.41)$$

According to equation (12.34) the second term on the right of equation (12.41) vanishes. In the first term the term yet to be evaluated can be done using table 12.1 (in atomic units):

$$\langle\psi_{2p_z}|\hat{z}|\psi_{2p_z}\rangle = \left(\frac{1}{4\sqrt{2\pi}}\right)^2 \int_0^{2\pi} d\varphi \int_0^\pi \sin(\theta) \cdot \cos^3(\theta) d\theta \int_0^\infty r^5 e^{-r} \, dr = 0 \quad (12.42)$$

which shows the first hyperpolarizability to be zero as expected.

12.3 The resolvent operator

A more elegant development of perturbation theory results from using concepts of section 7.3 to construct a projection operator known as the *resolvent*. In this approach we will make many of the same assumptions of section 12.1. There still exists an unperturbed ground state solution: $|0\rangle$, and ground state response functions to any order: $|\Phi_0^{(k)}\rangle$ can be expanded from a complementary set of zeroth-order eigenstates. Defining the ground state projection operator:

$$\hat{P}_0 = |0\rangle\langle 0| \qquad (12.43)$$

we revisit the derivation beginning with the intermediate normalization condition in equation (12.4). This is equivalent to:

$$\hat{P}_0|\Phi_0\rangle = |0\rangle \qquad (12.44)$$

Higher-order corrections are projected from $|\Phi_0\rangle$ using the zeroth-order orthogonal complement projection operator:

$$\hat{P}_n = \sum_{n\neq 0}|n\rangle\langle n| \qquad (12.45)$$

An arbitrary energy parameter: $\mathscr{E}$ is now added to the Schrödinger equation:

$$(\mathscr{E} + \hat{H})|\Phi_0\rangle = (\mathscr{E} + E_0)|\Phi_0\rangle \qquad (12.46)$$

In equation (12.46), $\hat{H}$, $|\Phi_0\rangle$, and E_0 are defined in equations (12.1), (12.3), and (12.6) respectively. Choosing the energy parameter to be: $\mathscr{E} = E_0$ leads to what is known as Brillouin–Wigner perturbation theory. If instead the parameter is set equal to the unperturbed ground state energy:

$$\mathscr{E} = E_0^{(0)} \qquad (12.47)$$

the development is then Rayleigh–Schrödinger perturbation theory from section 12.1, and will in fact give results having the same appearance we have already encountered. When this choice is made along with equation (12.1), we can rewrite equation (12.46) as:

$$\left(E_0^{(0)} - \hat{H}_0 \right) | \Phi_0 \rangle = \left(E_0^{(0)} - E_0 + \hat{V} \right) | \Phi_0 \rangle \tag{12.48}$$

equation (12.48) is now projected with the ground state orthogonal complement:

$$\hat{P}_n \left(E_0^{(0)} - \hat{H}_0 \right) | \Phi_0 \rangle = \hat{P}_n \left(E_0^{(0)} - E_0 + \hat{V} \right) | \Phi_0 \rangle \tag{12.49}$$

Operator $\hat{P}_n$ commutes with the zeroth-order energy and, because it is formed from its eigenfunctions, must also commute with the zeroth-order Hamiltonian. This allows us to rewrite equation (12.49) in the form:

$$\hat{P}_n | \Phi_0 \rangle = \left(E_0^{(0)} - \hat{H}_0 \right)^{-1} \hat{P}_n \left(E_0^{(0)} - E_0 + \hat{V} \right) | \Phi_0 \rangle \tag{12.50}$$

Equation (12.50) introduces the concept of an inverse operator. For instance, if there exists operator: $\hat{O}$ to eigenfunction: $| \psi_k \rangle$ with eigenvalue: o_k an inverse operator exhibits the property:

$$(c - \hat{O})^{-1} | \psi_k \rangle = \frac{1}{c - o_k} | \psi_k \rangle \tag{12.51}$$

Recall from section 7.3 that projection operators exhibit the idempotent property: $\hat{P}_n^2 = \hat{P}_n$. This condition is applied to equation (12.50), and the idempotent property produces a result written in the form:

$$\hat{P}_n | \Phi_0 \rangle = \hat{R}_{mn} \left(E_0^{(0)} - E_0 + \hat{V} \right) | \Phi_0 \rangle \tag{12.52}$$

Equation (12.52) introduces the *resolvent* operator:

$$\hat{R}_{mn} = \hat{P}_m \left(E_0^{(0)} - \hat{H}_0 \right)^{-1} \hat{P}_n = \sum_{m,n \neq 0} |m\rangle\langle m| \left(E_0^{(0)} - \hat{H}_0 \right)^{-1} |n\rangle\langle n| \tag{12.53}$$

The resolvent is also referred to as the *wave reaction operator*. The orthogonal complement is constructed from an orthonormal set of eigenfunctions to $\hat{H}_0$, so the bra-ket inverse operation demonstrates the following behavior:

$$\left\langle m \left| \left(E_0^{(0)} - \hat{H}_0 \right)^{-1} \right| n \right\rangle = \frac{1}{E_0^{(0)} - E_n^{(0)}} \langle m|n \rangle = \frac{1}{E_0^{(0)} - E_n^{(0)}} \delta_{mn} \tag{12.54}$$

which collapses one of the sums in equation (12.53), and simplifying it to the form:

$$\hat{R}_n = \sum_{n \neq 0} |n\rangle\langle n| \left(E_0^{(0)} - \hat{H}_0 \right)^{-1} |n\rangle\langle n| \tag{12.55}$$

As a consequence of either equation (12.53) or (12.55) the state: $| 0 \rangle$ must be non-degenerate to any member of its complement, otherwise the resolvent introduces a singularity into the expressions.

According to equations (12.3) and (12.13), the ground state response wave-function can be expressed in terms of the orthogonal complement projection operator (equation (12.45)) as:

$$|\Phi_0\rangle = |0\rangle + \left|\Phi_0^{(1)}\right\rangle + \left|\Phi_0^{(2)}\right\rangle + \ldots = |0\rangle + \hat{P}_n |\Phi_0\rangle \tag{12.56}$$

Equation (12.52) is directly inserted into the right-hand side of equation (12.56):

$$|\Phi_0\rangle = |0\rangle + \hat{R}_n\left(E_0^{(0)} - E_0 + \hat{V}\right) |\Phi_0\rangle \tag{12.57}$$

Iteration of equation (12.57) in the following fashion gives an infinite-order expansion of the ground state response function:

$$|\Phi_0\rangle = \sum_{k=0}^{\infty} \left[\hat{R}_{mn}\left(E_0^{(0)} - E_0 + \hat{V}\right)\right]^k |0\rangle \tag{12.58}$$

For instance, the first three terms are:

$$|\Phi_0\rangle = |0\rangle + \hat{R}_n \hat{V} |0\rangle + \hat{R}_n\left(E_0^{(0)} - E_0^{(1)} + \hat{V}\right)\hat{R}_n |0\rangle + \ldots$$

$$= |0\rangle + \sum_{n \neq 0} \frac{|n\rangle\langle n|\hat{V}|0\rangle}{E_0^{(0)} - E_n^{(0)}}$$

$$+ \left[\sum_{m,n \neq 0} \frac{|m\rangle\langle m|\hat{V}|n\rangle\langle n|\hat{V}|0\rangle}{\left(E_0^{(0)} - E_n^{(0)}\right)\left(E_0^{(0)} - E_m^{(0)}\right)} \right. \tag{12.59}$$

$$\left. - \left(E_0^{(1)} - E_0^{(0)}\right)\sum_{n \neq 0} \frac{|n\rangle\langle n|\hat{V}|0\rangle}{\left(E_0^{(0)} - E_n^{(0)}\right)^2} \right] + \ldots$$

When the perturbed Hamiltonian acts on this response function:

$$\hat{H} |\Phi_0\rangle = E_0 |\Phi_0\rangle \tag{12.60}$$

we left multiply equation (12.60) by: $\langle 0 |$ and with equation (12.1) find:

$$E_0 = E_0^{(0)} + \langle 0| \hat{V} |\Phi_0\rangle \tag{12.61}$$

Using the sequencing of an ordering parameter, the energy at any order $n > 0$ can be defined:

$$E_0^{(n)} - E_0^{(0)} = \left\langle 0 |\hat{V}| \Phi_0^{(n-1)} \right\rangle \tag{12.62}$$

into which equation (12.58) is inserted to obtain an order-by-order expansion that takes the same form as, for example, equation (12.18).

It is of interest to closely examine the third-order perturbed ground state energy:

$$E_0^{(3)} = \langle 0 |\hat{V}\hat{R}_n\hat{V}\hat{R}_m\hat{V}| 0\rangle - E_0^{(1)}\left\langle 0 \left|\hat{V}\hat{R}_n^2\hat{V}\right| 0\right\rangle \tag{12.63}$$

If this is expressed in terms of the first-order response function from equation (12.59) or (12.16) it can be rewritten as:

$$E_0^{(3)} = \left\langle \Phi_0^{(1)} \middle| \hat{V} \middle| \Phi_0^{(1)} \right\rangle - E_0^{(1)} \left\langle \Phi_0^{(1)} \middle| \Phi_0^{(1)} \right\rangle = \left\langle \Phi_0^{(1)} \middle| \hat{V} - E_0^{(1)} \middle| \Phi_0^{(1)} \right\rangle \qquad (12.64)$$

This is a result which can be generalized for odd-orders of Rayleigh–Schrödinger perturbation theory in which the kth order wavefunction can be used to determine the $(2k +1)$-order energy. For instance, we see from equation (12.64) the third-order energy from: $| \Phi_0^{(1)} \rangle$, and equation (12.9) shows the first-order energy obtained from: $| \Phi_0^{(0)} \rangle = | 0 \rangle$. The second term in the middle of equation (12.64) is called a *renormalization* or *exclusion principle violating* term, and is a consequence of the chosen intermediate normalization condition.

PARALLEL INVESTIGATION: Verify the fourth-order Rayleigh–Schrödinger perturbation expression is:

$$E_0^{(4)} = \sum_{k,m,n\neq 0} \frac{\langle 0 | \hat{V} | k \rangle \langle k | \hat{V} | m \rangle \langle m | \hat{V} | n \rangle \langle n | \hat{V} | 0 \rangle}{\left(E_0^{(0)} - E_k^{(0)}\right)\left(E_0^{(0)} - E_m^{(0)}\right)\left(E_0^{(0)} - E_n^{(0)}\right)} - E_0^{(2)} \sum_{n\neq 0} \frac{|\langle 0 | \hat{V} | n \rangle|^2}{\left(E_0^{(0)} - E_n^{(0)}\right)^2}$$

$$- 2E_0^{(1)} \sum_{m,n\neq 0} \frac{\langle 0 | \hat{V} | m \rangle \langle m | \hat{V} | n \rangle \langle n | \hat{V} | 0 \rangle}{\left(E_0^{(0)} - E_m^{(0)}\right)^2 \left(E_0^{(0)} - E_n^{(0)}\right)} + \left(E_0^{(1)}\right)^2 \sum_{n\neq 0} \frac{|\langle 0 | \hat{V} | n \rangle|^2}{\left(E_0^{(0)} - E_n^{(0)}\right)^3}$$

One last comment about the first-order perturbed wavefunction. Algorithms for its analytical determination were crucial to the efficacy of computational quantum chemistry. There was a point in time where higher order molecular properties were computationally determined by finding the ground state wavefunction relative to a Hamiltonian like equation (12.1), in which the perturbing term is explicitly included in the Hamiltonian in addition to any other potential in the zeroth-order Hamiltonian $\hat{H}_0$:

$$(\hat{H}_0 + \lambda \hat{V}) | \psi(\lambda) \rangle = E(\lambda) | \psi(\lambda) \rangle \qquad (12.65)$$

The energy is expanded in a power series:

$$E(\lambda) = E_0 + \frac{1}{1!}\left(\frac{\partial E(\lambda)}{\partial \lambda}\right) \cdot \lambda + \frac{1}{2!}\left(\frac{\partial^2 E(\lambda)}{\partial \lambda^2}\right) \cdot \lambda^2 + \cdots \qquad (12.66)$$

Properties at a particular order were equated to a *finite difference derivative*, for instance the first-order response energy would be found from:

$$E_1 = \left(\frac{\partial E(\lambda)}{\partial \lambda}\right) \approx \frac{E(\lambda_2) - E(\lambda_1)}{\lambda_2 - \lambda_1} \qquad (12.67)$$

A seminal point for computational chemistry came with the implementation of *analytical gradient* techniques These are based on the *Hellman–Feynman theorem*, which describes the way in which energy eigenvalue: E of a system with Hamiltonian: $\hat{H}$ and eigenfunction: $|\psi\rangle$ responds to perturbation: λ:

$$\frac{\partial E}{\partial \lambda} = \left\langle \frac{\partial \psi}{\partial \lambda} \middle| \hat{H} \middle| \psi \right\rangle + \left\langle \psi \middle| \frac{\partial \hat{H}}{\partial \lambda} \middle| \psi \right\rangle + \left\langle \psi \middle| \hat{H} \middle| \frac{\partial \psi}{\partial \lambda} \right\rangle \tag{12.68}$$

Comparison with equation (12.3) shows that:

$$\lim_{\lambda \to 0} \left| \frac{\partial \psi}{\partial \lambda} \right\rangle = \left| \Phi_0^{(1)} \right\rangle \tag{12.69}$$

so that analytical solution for $| \Phi_0^{(1)} \rangle$ via equation (12.16) becomes possible. A particular perturbation of importance is the electronic energy response to nuclear displacement. Implementation of such algorithms enable modern computational chemistry programs to determine equilibrium molecular geometries and transition states of concerted reactions.

12.4 Techniques for solving the sum over states equations

Before the advent of high speed computational methods, simplifications were made to facilitate solution of the Rayleigh–Schrödinger perturbation expressions. Among the more drastic is the average energy approximation, in which the energy difference denominators between the ground state and orthogonal complement: $E_0^{(0)} - E_n^{(0)}$ are replaced by a single average value. Under these conditions, equation (12.18) becomes:

$$E_0^{(2)}(ab) \approx \Delta E^{-1} \cdot \sum_{n \neq 0} \langle 0 | \hat{V}_a | n \rangle \langle n | \hat{V}_b | 0 \rangle \tag{12.70}$$

To further simplify equation (12.70) the ground state wavefunction $| 0 \rangle$ is introduced into the sum over states. This produces a projection operator of unique character:

$$\hat{1} = \sum_{n=0}^{\infty} | n \rangle \langle n | \tag{12.71}$$

Consider the action of this operator on any eigenstate of the orthonormal set:

$$\hat{1} | k \rangle = \sum_{n=0}^{\infty} | n \rangle \langle n | k \rangle = \sum_{n=0}^{\infty} | n \rangle \delta_{nk} = | k \rangle \tag{12.72}$$

Equation (12.72) demonstrates that a complete set of orthonormal functions can be used to form an identity operator. The sum in equation (12.71) is known as *resolution of the identity*. Equation (12.71) is inserted into equation (12.70) giving:

$$E_0^{(2)}(ab) \approx \Delta E^{-1} \cdot \langle 0 | \hat{V}_a \hat{1} \hat{V}_b | 0 \rangle \tag{12.73}$$

The summation in equation (12.71) contains the eigenstate: $| 0 \rangle$, so equation (12.72), shows that the identity operator: $\hat{1}$ has no effect on: $|0\rangle$. The commutation relation:

$[\hat{1}, \hat{V}] = 0$ therefore holds (to review commutation relations, see section 7.5), which allows us to write:

$$E_0^{(2)}(ab) \approx \Delta E^{-1} \cdot \langle 0 \,|\, \hat{V}_a \hat{1} \, \hat{V}_b \,|\, 0 \rangle = \Delta E^{-1} \cdot \langle 0 \,|\, \hat{V}_a \hat{V}_b \hat{1} \,|\, 0 \rangle$$
$$= \Delta E^{-1} \cdot \langle 0 \,|\, \hat{V}_a \hat{V}_b \,|\, 0 \rangle \qquad (12.74)$$

so that the second-order energy can be approximated as an expectation value over the ground state only. Taking this approach to determine the polarizability of hydrogen requires evaluation of the following integral:

$$\langle \psi_{1s} \,|\, z^2 \,|\, \psi_{1s} \rangle = \frac{1}{\pi} \int_0^{2\pi} d\varphi \int_0^{\pi} \sin(\theta) \cdot \cos^2(\theta) d\theta \int_0^{\infty} r^4 e^{-2r} \, dr = 1 \ \text{au} \qquad (12.75)$$

PARALLEL INVESTIGATION: Verify using the result of equation (12.70) that the experimental hydrogen atom polarizability of: $\alpha = 0.666 \ \text{Å}^3$ is reproduced using an average energy denominator: $\Delta E = -0.441$ au, which is 17.6% higher than the quantum mechanical difference between the $n = 1$ and $n = 2$ hydrogenic energy levels.

The average energy approximation is generally too drastic a simplification to be used if any meaningful conclusions are to be drawn from perturbation theory. Let us take some time to mention other approaches taken to solve the expressions. Ground state perturbation theory energies beyond first-order require the orthogonal complement: $|n\rangle$ to ground state wavefunction: $|0\rangle$ as a means of expanding response function: $|\Phi_0^{(k)}\rangle$. The set: $|n\rangle$ make it possible to form energy denominators required for all expressions beyond first order in a relatively straightforward manner. This is because elements of the orthogonal complement: (i) are eigenfunctions of the zeroth-order Hamiltonian: $\hat{H}_0$, and (ii) form an orthonormal set.

With these conditions in mind, we can therefore represent the sum over states expressions in a matrix form. For the second-order energy, we write:

$$E_0^{(2)}(ab) = \mathbf{V}_\mathbf{a}^\mathbf{T} \mathbf{E}^{-1} \mathbf{V}_\mathbf{b} \qquad (12.76)$$

If the orthogonal complement includes n eigenstates, then $\mathbf{V}_\mathbf{b}$ is an $n \times 1$ column matrix of elements: $\langle n \,|\, \hat{V}_b \,|\, 0 \rangle$, $\mathbf{V}_\mathbf{a}^\mathbf{T}$ is a $1 \times n$ row matrix of elements: $\langle 0 \,|\, \hat{V}_a \,|\, n \rangle$, and $\mathbf{E}^{-1}$ is an $n \times n$ diagonal matrix with elements: $1/(E_0^{(0)} - E_n^{(0)})$. To review properties of matrices, see section 7.4. In general, we should use Hermitian conjugate rather than transposed vectors, but for real eigenstates: $\mathbf{V}_\mathbf{a}^\dagger = \mathbf{V}_\mathbf{a}^\mathbf{T}$. In equation (12.76), the conforming matrices of the triple product produce a $(1 \times n) \times (n \times n) \times (n \times 1) = 1 \times 1$ or numeric result. In a similar fashion the third-order perturbation energy has matrix form:

$$E_0^{(3)}(abc) = \mathbf{V}_\mathbf{a}^\mathbf{T} \mathbf{E}^{-1} \overline{\mathbf{V}}_\mathbf{b} \mathbf{E}^{-1} \mathbf{V}_\mathbf{c} - E_0^{(1)}(b) \cdot \mathbf{V}_\mathbf{a}^\mathbf{T} \mathbf{E}^{-2} \mathbf{V}_\mathbf{c} \qquad (12.77)$$

Equation (12.77) uses many of the same conventions of equation (12.76). The matrix product of the second term is multiplied by the scalar value: $E_0^{(1)}(b) = \langle 0 \,|\, \hat{V}_b \,|\, 0 \rangle$, and also involves a diagonal matrix of the square difference of energy eigenvalues: $1/(E_0^{(0)} - E_n^{(0)})^2$. The first term includes: $\overline{V}_b$, an $n \times n$ square matrix with elements: $\langle m | \hat{V}_b | n \rangle$. Notice both matrix products in equation (12.77) conform so that the overall result again is a 1×1 or numeric answer.

It is also possible that energy denominators may be constructed from a set of functions which couple across the zeroth-order Hamiltonian, forming a non-diagonal energy representation. In this case, the techniques of chapter 10 can be employed to first diagonalize the energy representation. In the following we assume these matrix elements to be formed from real wavefunctions, so that the energy representation is a symmetric rather than Hermitian matrix. According to section 10.3, diagonalizing this matrix is accomplished by an orthogonal transformation:

$$e = C^T E \, C \qquad (12.78)$$

The energy denominator: e^{-1} can then be formed by inverting the diagonal elements of: e. It is then required that the following transformations be performed on the column vector:

$$v_b = C^T V_b \qquad (12.79)$$

and the row vector:

$$v_a^T = V_a^T C \qquad (12.80)$$

These matrices are then used to evaluate the second-order energy in the exact same way as equation (12.76):

$$E_0^{(2)}(ab) = v_a^T e^{-1} v_b \qquad (12.81)$$

An interesting result occurs upon applying properties of the inverse of a matrix product to equation (12.81). Suppose we perform a multiplication of three matrices $A \, B$ and C. The inverse of this product is:

$$(A \, B \, C)^{-1} = C^{-1} B^{-1} A^{-1} \qquad (12.82)$$

This can be argued from the fact that the inverse operation of putting on your socks and shoes is taking off your shoes and socks, or can be proved in a straightforward fashion by successive multiplication on the matrix product: $P = A \, B \, C$:

$$P^{-1} P = C^{-1} B^{-1} A^{-1} P \; = C^{-1} B^{-1} A^{-1} A \, B \, C = C^{-1} B^{-1} B \, C = C^{-1} C = 1 \quad (12.83)$$

Equation (12.78) is inserted into equation (12.81), and we use equation (12.82) to give:

$$e^{-1} = (C^T E \, C)^{-1} = C^{-1} E^{-1} (C^T)^{-1} = C^T E^{-1} C \qquad (12.84)$$

The final identity in equation (12.84) follows from properties of orthogonal transformation matrices (see section 10.3)). equations (12.84), (12.79), and (12.80) are used in equation (12.81) giving:

$$E_0^{(2)}(ab) = v_a^T e^{-1} v_b = V_a^T C \, C^T E^{-1} C \, C^T V_b = V_a^T E^{-1} V_b \qquad (12.85)$$

> **PARALLEL INVESTIGATION:** Verify using the properties of orthogonal transformation that the third-order Rayleigh–Schrödinger perturbation theory energy can be expressed in equivalent diagonal and non-diagonal forms:
> $$E_0^{(3)}(abc) = \mathbf{v}_a^T \mathbf{e}^{-1} \bar{\mathbf{v}}_b \mathbf{e}^{-1} \mathbf{v}_c - E_0^{(1)}(b) \cdot \mathbf{v}_a^T \mathbf{e}^{-2} \mathbf{v}_c = \mathbf{V}_a^T \mathbf{E}^{-1} \bar{\mathbf{V}}_b \mathbf{E}^{-1} \mathbf{V}_c - E_0^{(1)}(b) \cdot \mathbf{V}_a^T \mathbf{E}^{-2} \mathbf{V}_c$$

Equation (12.85) shows how a sum over states expression can be evaluated from a basis of orthonormal eigenstates which are not eigenfunctions of the zeroth-order Hamiltonian by matrix inversion rather than matrix diagonalization. A process such as this is referred to as a direct method.

From linear algebra considerations, a non-diagonal solution of the sum over states expression may not seem to have any advantage over the diagonal form. Although a few transformation steps in equations (12.79) and (12.80) are avoided, the non-diagonal solution still requires a matrix inversion, which in computational terms is about the same computer time and space as a diagonalization. These issues become prohibitive as the number of states in the orthogonal complement grow substantially in an attempt to correctly represent the energy differences between states. However, an algorithm can be applied to the matrix inversion problem which makes it an overwhelmingly better option. This process is sometimes referred to as the reduced linear equations method, but is more aptly called *direct inversion in the iterative subspace*.

Suppose there exists a non-diagonal $n \times n$ symmetric matrix: $\mathbf{X}$ and a known $n \times 1$ column matrix: $\mathbf{Y}$ which are required in an expression of the type:

$$\mathbf{X}^{-1}\mathbf{Y} = \mathbf{Z} \tag{12.86}$$

as is required in equation (12.85). The task of interest is to find the $n \times 1$ column product matrix: $\mathbf{Z}$, which can be determined without knowledge of the inverse matrix if the following system of equations can be solved:

$$\mathbf{X}\,\mathbf{Z} = \mathbf{Y} \tag{12.87}$$

Column $\mathbf{Z}$ is approximated as a linear combination of columns:

$$\mathbf{Z} = \sum_i a_i \mathbf{z}_i \tag{12.88}$$

with the matrix product definition:

$$\mathbf{X}\,\mathbf{z}_i = \mathbf{b}_i \tag{12.89}$$

so that:

$$\mathbf{X}\,\mathbf{Z} = \sum_i a_i \mathbf{b}_i \tag{12.90}$$

Upon substituting equation (12.90) into equation (12.87), an expression for the variance of the approximate solution is obtained:

$$\delta^2 = \left(\mathbf{Y} - \sum_i a_i \mathbf{b}_i \right)^2 = \mathbf{Y}^T \mathbf{Y} - 2 \sum_i \mathbf{Y}^T \cdot a_i \mathbf{b}_i + \sum_{i,j} \mathbf{b}_i^T \mathbf{b}_j a_i a_j \qquad (12.91)$$

Notice that the matrix multiplications in equation (12.91) are scalar products of rows × columns.

The variance is minimized with respect to the column expansion coefficients:

$$\frac{\partial(\delta^2)}{\partial a_k} = 0 = -2\mathbf{Y}^T \mathbf{b} + 2 \sum_i a_i \mathbf{b}_i^T \mathbf{b}_k \qquad (12.92)$$

Using the definitions:

$$(\mathbf{P})_{ik} = \mathbf{b}_i^T \mathbf{b}_k \quad \mathbf{Q}_k = \mathbf{Y}^T \mathbf{b}_k \qquad (12.93)$$

the equation for expansion coefficients: a_i is written in matrix form:

$$\mathbf{P a} = \mathbf{Q} \qquad (12.94)$$

The best-fit coefficients are then determined from the matrix problem:

$$\mathbf{a} = \mathbf{P}^{-1} \mathbf{Q} \qquad (12.95)$$

It appears from equations (12.95) and (12.86) that one inversion problem has simply been replaced by another. However, the dimensionality of equation (12.95) is that of the number of expansion columns, rather than the dimension of the orthogonal complement.

An excellent starting guess for the first expansion column is provided by dividing column vector $\mathbf{Y}$ by diagonal elements of the inverse energy representation:

$$\mathbf{z}_1(i) = (\mathbf{X})_{ii}^{-1} \cdot \mathbf{Y}_i \qquad (12.96)$$

This is particularly effective because the matrix $\mathbf{X}$ is typically very diagonal dominant. At each iteration, the value: δ^2 from equation (12.91) is compared against some predetermined tolerance. If this value is not satisfied, then the next expansion vector: $\mathbf{z}_k$ is supplied from its residuum:

$$\mathbf{z}_k = \mathbf{Y} - \sum_{i<k} a_i \mathbf{b}_i \qquad (12.97)$$

During the process, the expansion vectors can be Schmidt orthogonalized (see section 7.3) to eliminate linear dependencies. Typically, the process successfully converges after a handful of iterations, so that the largest matrix to diagonalize is order 10 × 10 or less. This is substantially smaller than the Terabytes of space required in some instances.

What's the Matter with Waves?

An introduction to techniques and applications of quantum mechanics

William Parkinson

Chapter 13

Electrons in molecules

13.1 The simplest molecular model: a one-electron diatomic

Section 11.1 is a watershed of sorts. It details what may be argued to be the pinnacle of success in quantum mechanics—analytical solutions of the one electron hydrogen atom Hamiltonian. The excitement has waned by section 11.5, where it is learned that introducing even one more electron to the central potential problem requires either drastic approximate or numerical solutions of non-linear equations. In this chapter, the complexity of a system is compounded by adding additional nuclear centers to the many electron problem. We must again face the unsatisfactory fact that major approximations and simplifications are needed to obtain solutions.

To appreciate the challenges posed by a molecule, consider the simple combination of atoms depicted in figure 13.1. This represents a heteronuclear diatomic (nuclei labeled A and B), with two electrons (labeled 1 and 2) bonding them together. The Hamiltonian for the four particles is:

$$\hat{H} = -\frac{\hbar^2}{2M_A}\hat{\nabla}_A^2 - \frac{\hbar^2}{2M_B}\hat{\nabla}_B^2 - \frac{\hbar^2}{2m_e}\hat{\nabla}_1^2 - \frac{\hbar^2}{2m_e}\hat{\nabla}_2^2$$
$$- \frac{Z_A e^2}{4\pi\varepsilon_0 r_{1A}} - \frac{Z_A e^2}{4\pi\varepsilon_0 r_{2A}} - \frac{Z_B e^2}{4\pi\varepsilon_0 r_{1B}} - \frac{Z_B e^2}{4\pi\varepsilon_0 r_{2B}} + \frac{Z_A Z_B e^2}{4\pi\varepsilon_0 R_{AB}} + \frac{e^2}{4\pi\varepsilon_0 r_{12}} \tag{13.1}$$

Equation (13.1) has four negative potential terms describing attraction between electrons and nuclei. There are also two positive terms representing nuclear–nuclear and electron–electron repulsion. When used in an eigenvalue problem, equation (13.1) produces an untractable differential equation, with complications additional to those encountered in section 11.5. As we inevitably faced then, analytically solving equation (13.1) requires assumptions, simplifications, or outright omissions.

As a first step towards rendering a solvable Hamiltonian, we employ the *Born–Oppenheimer approximation*, which uses the large mass difference between electrons and nuclei to allow neglect of nuclear kinetic energy terms. In other words, the electrons instantaneously adjust to the nuclear positions. Along with eliminating the

doi:10.1088/978-1-6817-4577-0ch13

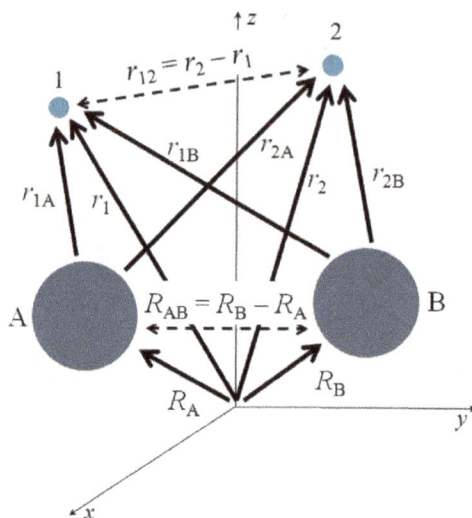

Figure 13.1. Coordinate system for a heteronuclear diatomic molecule.

first two terms in equation (13.1), this also makes R_{AB} a fixed value, this term is thus a constant depending on a chosen location of the nuclei.

This is not enough to make equation (13.1) analytically solvable. We will simplify matters even more by binding nuclei A and B with a single electron. In addition to terms eliminated by the Born–Oppenheimer approximation, all terms referencing electron 2 along with the electron–electron repulsion also vanish:

$$\hat{H} = \hat{H}_1 + V_{AB} \tag{13.2}$$

The constant V_{AB} represents parametric nuclear repulsion. With these choices the Schrödinger equation is:

$$\hat{H}\Psi = (\hat{H}_1 + V_{AB})\Psi = E\Psi + V_{AB}\Psi = (E + V_{AB})\Psi \tag{13.3}$$

The wavefunctions: Ψ are eigenfunctions of Hamiltonian: $\hat{H}$ with eigenvalues: E that parametrically depend on the location of A and B through electron-nuclear attraction terms, but not on the repulsion between A and B.

In other words, eigenvalue E is a scaled by constant: V_{AB}. In fact, E is the same eigenvalue and Ψ the same eigenfunction whether or not V_{AB} is added. This is apparent by comparing to the un-scaled eigenvalue equation:

$$\hat{H}\Psi = E\Psi \tag{13.4}$$

Adding: $V_{AB}\Psi$ to each side of equation (13.4) gives equation (13.3) only if both E and Ψ are the same in both equations. We can therefore simply solve the one-electron Hamiltonian equation for its energy then add in the nuclear repulsion term afterward.

A solution is proposed to equation (13.3) which uses a weighted combination of a single function centered on each nucleus:

$$\Psi = c_A \varphi_A + c_B \varphi_B \tag{13.5}$$

Equation (13.5) uses a *linear combination of atomic orbitals*. The function Ψ is referred to as a *molecular orbital*. Collectively, the process is the LCAO MO approximation. Equations (13.2) and (13.5) are used with the variational theorem (see section 10.3) to determine the best fit coefficients through a secular equation:

$$(\mathbf{H} - \mathbf{ES})\mathbf{C} = \mathbf{0}. \tag{13.6}$$

Since our molecular orbital is an expansion given by equation (13.5), $\mathbf{H}$ and $\mathbf{S}$ are therefore symmetric 2×2 matrices representing the Hamiltonian and overlap respectively, $\mathbf{E}$ is a 2×2 matrix with energy eigenvalues along the diagonal, and $\mathbf{C}$ is a 2×2 matrix of the best fit coefficients. What follows are three separate approaches to solving this equation using either different assumptions or different methodologies. Each lends its own perspective to the problem, and gives insight to a general picture of the molecular orbital approach.

Case 1: Solution of the secular determinant ignoring overlap. Using techniques from section 10.3, we first note left-multiplying both sides of equation (13.6) by $(\mathbf{H} - \mathbf{ES})^{-1}$ gives the trivial solution: $\mathbf{C} = \mathbf{0}$. To prevent this, $(\mathbf{H} - \mathbf{ES})$ must have no inverse, so therefore according to linear algebra has zero determinant:

$$\det(\mathbf{H} - \mathbf{ES}) = \det \begin{vmatrix} \alpha_A - E & \beta_{AB} - ES_{AB} \\ \beta_{AB} - ES_{AB} & \alpha_B - E \end{vmatrix} = 0 \tag{13.7}$$

Matrix element: S_{AB} is an overlap integral introduced in section 7.2:

$$S_{AB} = \langle \varphi_A | \varphi_B \rangle \tag{13.8}$$

At first it might appear that orthonormality of basis functions means: $S_{AB} = \delta_{AB}$ (the Kronecker delta, see section 7.2), however keep in mind these orbitals are located on different nuclei. This is a *two-center integral*. The value: α_A is defined:

$$\alpha_A = \langle \varphi_A | \hat{H} | \varphi_A \rangle \tag{13.9}$$

with a similar definition for α_B. The value β_{AB} is referred to as a *resonance integral*:

$$\beta_{AB} = \langle \varphi_A | \hat{H} | \varphi_B \rangle \tag{13.10}$$

As is in the case of overlap, resonance is a two-center integral.

Equation (13.7) is expanded and terms are collected in powers of E:

$$\begin{aligned}(\alpha_A - E)(\alpha_B - E) - (\beta_{AB} - ES_{AB})^2 &= 0 \\ = E^2(1 - S_{AB}^2) + (2\beta_{AB}S_{AB} - \alpha_A - \alpha_B)E + \left(\alpha_A\alpha_B - \beta_{AB}^2\right)\end{aligned} \tag{13.11}$$

Equation (13.11) is a second-order polynomial in E, therefore its coefficients can be inserted into the quadratic formula to find its roots. After cancellation of a few terms we obtain:

$$E = \frac{(\alpha_A + \alpha_B - 2\beta_{AB})}{2(1 - S_{AB}^2)} \pm$$
$$\frac{\sqrt{(\alpha_A - \alpha_B)^2 + 4\beta_{AB}^2 + 4S_{AB}^2\alpha_A\alpha_B - 4\beta_{AB}S_{AB}(\alpha_A + \alpha_B)}}{2\left(1 - S_{AB}^2\right)} \tag{13.12}$$

If overlap is neglected, the roots of equation (13.12) simplify to:

$$E = \frac{\alpha_A + \alpha_B}{2} \pm \frac{\sqrt{(\alpha_A - \alpha_B)^2 + 4\beta_{AB}^2}}{2} \tag{13.13}$$

Figure 13.2 diagrammatically expresses this result. We see from the definition in equation (13.9) that α_A and α_B represent the *self-energies* of orbitals centered on A and B respectively. The first term of equation (13.13) is therefore the average of these energies. The second term in equation (13.13) gives an amount by which the average energy is either decreased or increased when atoms are combined in the molecular state.

For further interpretation, equation (13.13) is expressed in the form:

$$E = \frac{\alpha_A + \alpha_B}{2} \pm \frac{(\alpha_A - \alpha_B)}{2} \cdot \sqrt{1 + x} \tag{13.14}$$

with:

$$x = \frac{4\beta_{AB}^2}{(\alpha_A - \alpha_B)^2} \tag{13.15}$$

The square root is expanded in a Taylor series:

$$\sqrt{1 + x} = 1 + \frac{1}{2}x - \frac{1}{8}x^2 + \dots \tag{13.16}$$

which is inserted into equation (13.14):

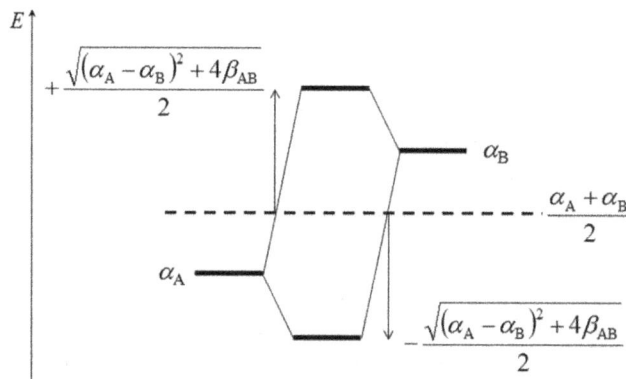

Figure 13.2. Molecular orbital energy diagram for a one-electron heteronuclear diatomic with zero overlap.

$$E = \frac{\alpha_A + \alpha_B}{2} \pm \frac{(\alpha_A - \alpha_B)}{2} \cdot \left(1 + \frac{1}{2}\frac{4\beta_{AB}^2}{(\alpha_A - \alpha_B)^2} - \ldots\right) \tag{13.17}$$

The energies of the molecular orbitals are then:

$$E_1 = \alpha_A + \frac{\beta_{AB}^2}{\alpha_A - \alpha_B} - \ldots$$

$$E_2 = \alpha_B - \frac{\beta_{AB}^2}{\alpha_A - \alpha_B} + \ldots \tag{13.18}$$

Arbitrarily making the assumption that: $\alpha_A < \alpha_B$, eigenvalue E_1 is not only lower than the average energy, it is in fact lower than α_A. Occupying this orbital stabilizes the molecular energy relative to either separated atom. If $\alpha_A < \alpha_B$, this also shows E_2 to be higher in energy than α_B, but this choice can be reversed with no loss of consistency. Using $\alpha_B < \alpha_A$ now makes E_1 higher in energy than α_A, and E_2 lower than α_B. The molecular orbital pushed below the separated atoms is a *bonding orbital*. The one higher in energy is destabilizing relative to the separated atoms, and is therefore named an *antibonding orbital*.

Equation (13.18) also illustrates a key feature to interpreting the degree of *interaction* of atomic orbitals as they combine to form a molecular orbital. According to equation (13.18), the energy shift is a function of resonance integral: β_{AB}, thus this value in some way quantifies atomic interaction. In addition, the denominators of equation (13.18) show as $\alpha_A \to \alpha_B$, the degree of downward and upward shift in bonding and antibonding orbital energies increases. This implies the proximity of atomic orbital energy is inherently related to their interaction in forming molecular orbitals. Of course, if they become too close in energy, the denominators become too small, and the Taylor series expansion is invalid. That aside, it is generally true that atomic orbitals close in energy (and also having the proper symmetry) interact most effectively in molecular orbital formation. This concept forms the essence of what is known by chemists as *ligand field theory*. An energy diagram of molecular orbitals resulting from the atomic orbital interaction picture is depicted in figure 13.2.

PARALLEL INVESTIGATION: Verify the following for a homonuclear triatomic molecule with three bonded atoms at the corners of an equilateral triangle. Given that each atom contributes one atomic orbital to the molecule, set up and symbolically solve the secular determinant using the simplifications: (1) S values are zero, (2) all α values are equal, and (3) all β values are equal. Use the substitution: $x = (\alpha - E)/\beta$ to write the secular determinant in the form: $x^3 - 3x + 2 = 0$. Factor this cubic polynomial to find the orbital energies of the homonuclear triatomic are: $E_1 = \alpha + 2\beta$, $E_2 = \alpha - \beta$, and $E_3 = \alpha - \beta$.

Case 2. Solving the secular determinant for a homonuclear diatomic including overlap.
Let us return to equation (13.7), but this time using a homonuclear system with: $\alpha_A = \alpha_B = \alpha$. The secular determinant becomes:

$$\det \begin{vmatrix} \alpha - E & \beta_{AB} - ES_{AB} \\ \beta_{AB} - ES_{AB} & \alpha - E \end{vmatrix} = 0 = (\alpha - E)^2 - (\beta_{AB} - ES_{AB})^2 \qquad (13.19)$$

Rearranging and taking the square root of both sides gives:

$$=(\alpha - E) = \pm(\beta_{AB} - ES_{AB}) \qquad (13.20)$$

When terms in E are gathered, we obtain two eigenvalues:

$$E_1 = \frac{\alpha + \beta_{AB}}{1 + S_{AB}}$$

$$E_2 = \frac{\alpha - \beta_{AB}}{1 - S_{AB}} \qquad (13.21)$$

Given that integral: β_{AB} is a negative quantity, E_1 is then a bonding orbital with energy below: α and E_2 is an antibonding orbital raised above: α. Making the reasonable assumption that ϕ are normalized functions, the two-center overlap thus is restricted to values: $0 \leqslant S_{AB} \leqslant 1$, as the atomic orbitals on each center go from no to maximum coincidence. Including overlap, the denominators of equation (13.21) show that E_2 is raised in energy to a greater extent than E_1 is lowered. This implies *the antibonding orbital is more antibonding than the bonding orbital is bonding.*

Case 3. Optimizing the linear combination via matrix diagonalization.
Suppose we now minimize the energy of the linear combination by Hamiltonian matrix diagonalization. For simplicity, we again take the homonuclear case with no overlap of atomic orbitals:

$$\mathbf{H} = \begin{vmatrix} \alpha - E & \beta_{AB} \\ \beta_{AB} & \alpha - E \end{vmatrix} \qquad (13.22)$$

Matrix elements of equation (13.22) are divided by: β_{AB}. We then use definitions in the form:

$$x_1 = -\frac{(\alpha - E_1)}{\beta_{AB}} \qquad (13.23)$$

to write:

$$\mathbf{H} = \begin{vmatrix} -x_1 & 1 \\ 1 & -x_2 \end{vmatrix} \qquad (13.24)$$

Equation (13.24) is inserted into equation (13.6). A diagonal matrix of x values is factored out, and moved to the other side of the matrix equation:

$$\mathbf{H'C = XC} \tag{13.25}$$

Equation (13.25) uses the definition:

$$\mathbf{H'} = \begin{vmatrix} 0 & 1 \\ 1 & 0 \end{vmatrix} \tag{13.26}$$

Equation (13.25) is diagonalized by orthogonal transformation matrix $\mathbf{C}$. Doing this we find:

$$\mathbf{X} = \begin{vmatrix} 1 & 0 \\ 0 & -1 \end{vmatrix}$$

$$\mathbf{C} = \begin{vmatrix} 0.707 & 0.707 \\ 0.707 & -0.707 \end{vmatrix} \tag{13.27}$$

Using the diagonal elements of the $\mathbf{X}$ matrix expressed in the form of equation (13.23), we obtain the eigenvalues: $E_1 = \alpha + \beta_{AB}$ and $E_2 = \alpha - \beta_{AB}$, the same values in equation (13.21) when S_{AB} is set to zero. Because $\mathbf{H'}$ is symmetric, $\mathbf{C}$ is an orthogonal transformation matrix: $\mathbf{C}^T\mathbf{C} = \mathbf{1}$, with two orthonormal column vectors. These are used with basis functions: ϕ_i to form a bonding and antibonding molecular orbital:

$$\Psi_1 = 0.707\varphi_A + 0.707\varphi_B = \frac{1}{\sqrt{2}}(\varphi_A + \varphi_B)$$

$$\Psi_2 = 0.707\varphi_A - 0.707\varphi_B = \frac{1}{\sqrt{2}}(\varphi_A - \varphi_B) \tag{13.28}$$

PARALLEL INVESTIGATION: Verify the following for a homonuclear triatomic molecule with three atoms bonded at the corners of an equilateral triangle. Set up the 3 × 3 Hamiltonian matrix using the simplifications: (1) S values are zero, (2) all α values are equal, and (3) all β values are equal. Following equations (13.22)–(13.28), show the diagonalization method gives the same eigenvalues as the secular determinant factorization, and that the three molecular orbital wavefunctions are: $\Psi_1 = 0.577\varphi_1 + 0.577\varphi_2 + 0.577\varphi_3$, $\Psi_2 = 0.638\varphi_1 - 0.760\varphi_2 + 0.122\varphi_3$, and $\Psi_3 = 0.509\varphi_1 + 0.298\varphi_2 - 0.807\varphi_3$.

In theory, the coefficients have values which can possibly range between: $0 \leqslant c \leqslant 1$, as an orbital goes from no interaction to what can be described as the ionic case where electron preferentially transferred to one atom over the other. As may be anticipated for a homonuclear diatomic, the molecular orbitals are an equally weighted combination of a basis function from each nuclear center. Bonding orbital: Ψ_1 is a constructive wave superposition and antibonding orbital: Ψ_2 a destructive superposition. Figure 13.3 shows a 2-dimensional amplitude plot for bonding molecular orbital: Ψ_1, by using hydrogenic 1s orbitals with $Z = 1$ (see table 11.1) for basis functions: ϕ_A and ϕ_B. As nuclear separation decreases from 6.0 Å to the

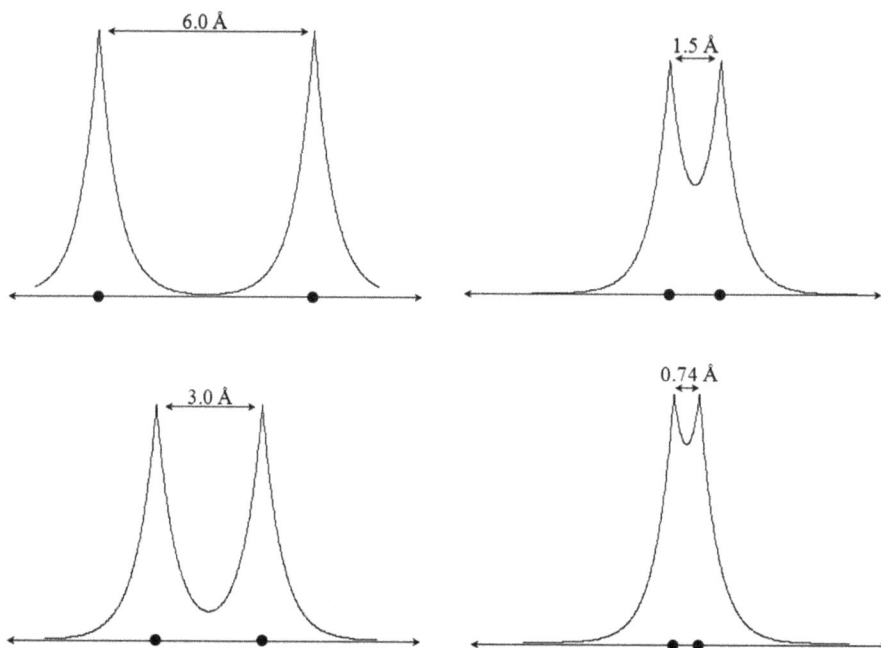

Figure 13.3. Amplitude cross-section of H_2 bonding molecular orbital: $\Psi_1 = 0.707\phi_A + 0.707\phi_B$ as a function of internuclear distance.

experimental H_2 molecular bond length of 0.74 Å, there is distinct enhancement of wave amplitude in the region between hydrogen atoms. Recalling that a particle's location is determined from the square amplitude, it is apparent that electrons in bonding orbitals have enhanced likelihood of being between the nuclei, thereby shielding their positive charge, and allowing a nuclear separation of 0.74 Å. Figure 13.4(a) gives a 2-dimensional slice of amplitude for antibonding molecular orbital: Ψ_2. The phase difference between atomic orbitals creates a node in the bonding region. The square amplitude plot of figure 13.4 (b) demonstrates the lack of electron density between nuclei, resulting in nuclear repulsion and molecular destabilization.

According to equation (13.28) the homonuclear bonding molecular orbital can be expressed in the form:

$$\Psi_1 = N(\varphi_A + \varphi_B) \tag{13.29}$$

where N is a normalizing factor. Assuming the ϕ are real and normalized, we obtain:

$$\langle \Psi_1 | \Psi_1 \rangle = 1 = N^2(\langle \varphi_A | \varphi_A \rangle + \langle \varphi_B | \varphi_B \rangle + 2\langle \varphi_A | \varphi_B \rangle) = N^2(2 + 2S_{AB}) \tag{13.30}$$

The bonding molecular orbital thus has normalization factor:

$$N = \frac{1}{(2 + 2S_{AB})^{1/2}} \tag{13.31}$$

Beginning with the antibonding orbital in the form: $\Psi_2 = N'(\varphi_A - \varphi_B)$ and proceeding in a similar fashion, we obtain:

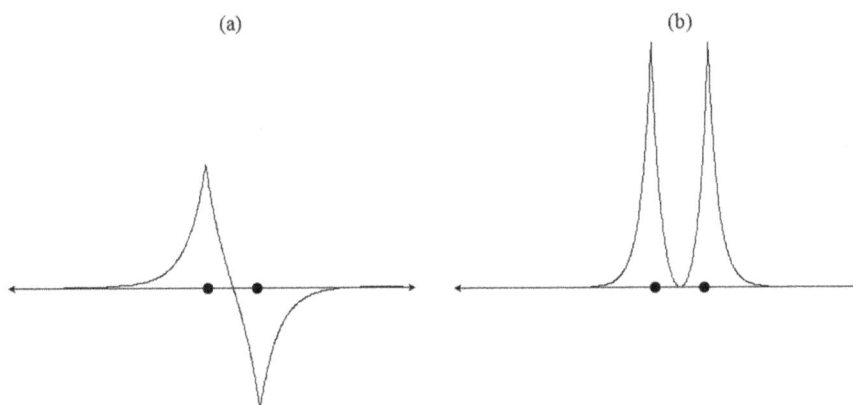

Figure 13.4. (a). Amplitude cross-section of H_2 antibonding molecular orbital. (b). Probability density cross-section of H_2 antibonding molecular orbital.

$$N' = \frac{1}{(2-2S_{AB})^{1/2}} \tag{13.32}$$

If the two-center overlap is chosen: $S_{AB} = 0$ as stipulated in case 3, the normalizing factors in equations (13.31) and (13.32) are the exact values determined from the variational solution shown in equation (13.28).

PARALLEL INVESTIGATION: Verify the following for a homonuclear triatomic molecule with three atoms bonded at the corners of an equilateral triangle. Given that each atom contributes a single atomic orbital, the bonding molecular orbital given by: $\Psi = N(\varphi_1 + \varphi_2 + \varphi_3)$ has normalizing factor: $N = \dfrac{1}{(3 + 2S_{AB} + 2S_{AC} + 2S_{BC})^{1/2}}$, and that this gives the same ground state wavefunction expansion coefficients as the variational treatment provided that overlap is ignored.

13.2 The hydrogen molecule

Let us now use knowledge gained in section 13.1 to discuss a two-electron hydrogen molecule. Beginning with equation (13.1), we again invoke the Born–Oppenheimer approximation. To further facilitate its solution, we ignore the electron–electron repulsion term. Equation (13.1) now simplifies to a separable electronic Hamiltonian of non-interacting particles:

$$\hat{H} = \hat{H}_1 + \hat{H}_2 + V_{AB} \tag{13.33}$$

Hamiltonian: $\hat{H}_1$ contains the kinetic energy and nuclear attraction for electron 1 only, while $\hat{H}_2$ expresses similar terms for electron 2. As was the case in section 13.1, nuclear repulsion: V_{AB} merely scales the total energy as a parametric addition to the electronic energy.

Because the Hamiltonian is separable, the two-electron wavefunction can be expressed as a product of two one-electron homonuclear diatomic cation wavefunctions, like those described in section 13.1. Using equation (13.29) for the bonding molecular orbital, we simply employ two hydrogenic 1s orbitals as basis functions:

$$\Psi_1 = \frac{1}{\sqrt{2}}(1s_A + 1s_B) \tag{13.34}$$

The electronic wavefunction composed of two fermions must be properly anti-symmetrized. This requires a Slater determinant (see section 11.6) to represent the H_2 molecular wavefunction:

$$\Phi_{H_2}$$

$$= \frac{1}{\sqrt{2}} \det \begin{vmatrix} \Psi_1(1)\alpha(1) & \Psi_1(1)\beta(1) \\ \Psi_1(2)\alpha(2) & \Psi_1(2)\beta(2) \end{vmatrix} = \frac{1}{\sqrt{2}}\Psi_1(1)\Psi_1(2)[\alpha(1)\beta(2) - \beta(1)\alpha(2)] \tag{13.35}$$

The factor: $1/\sqrt{2}$ in equation (13.35) normalizes the spin function in addition to the normalizer included for Ψ_1 in equation (13.34).

A problem with physical interpretation of the spatial wavefunction of equation (13.35) is illuminated by expanding it and inserting equation (13.34):

$$\Psi_1(1)\Psi_2(2)$$

$$= \frac{1}{2}[1s_A(1)1s_A(2) + 1s_A(1)1s_B(2) + 1s_B(1)1s_A(2) + 1s_B(1)1s_B(2)] \tag{13.36}$$

The first and last terms of equation (13.36) place both electrons on either center A or B, suggesting half the time H_2 dissociates should produce H^+ (a proton) and H^-. It is known experimentally that molecular hydrogen dissociates into two neutral hydrogen atoms. Wavefunctions which incorrectly represent dissociation are said to lack *size consistency*.

One remedy to this situation is *valence bond theory* proposed by Heitler and London. This representation, which actually predates molecular orbital theory, is again a Born–Oppenheimer, non-interacting electron picture. Instead of a weighted sum of one-electron functions like molecular orbital theory, the valence bond wavefunction is a product of functions from each center. To keep the electrons indistinguishable, the electron labels must be interchanged with a second product in the spatial wavefunction. Including two-electron antisymmetry gives the following valence bond wavefunction for the hydrogen molecule:

$$\Phi_{H_2} = N[1s_A(1)1s_B(2) + 1s_B(1)1s_A(2)]\cdot[\alpha(1)\beta(2) - \beta(1)\alpha(2)] \tag{13.37}$$

When equation (13.37) is expanded and terms are collected, it can be expressed in the form of two Slater determinants:

$$\Phi_{H_2} = N \det \begin{vmatrix} 1s_A(1)\alpha(1) & 1s_B(1)\beta(1) \\ 1s_A(2)\alpha(2) & 1s_B(2)\beta(2) \end{vmatrix} - N \det \begin{vmatrix} 1s_A(1)\beta(1) & 1s_B(1)\alpha(1) \\ 1s_A(2)\beta(2) & 1s_B(2)\alpha(2) \end{vmatrix} \tag{13.38}$$

A key feature of equation (13.38) is a correct dissociation of H_2 into two neutral H atoms. Equation (13.38) is known as a multi-determinant *multi-configuration* or *multi-reference* ground state electronic representation. Equation (13.35) uses a single determinant, single configuration, or single reference ground state. The multi-reference issue is also important in correctly representing atomic systems such as beryllium, which cannot be successfully described without inclusion of 2p along with 2s character in the ground state wavefunction.

13.3 Practical information regarding calculations

In computational applications, valence bond methods and multi-reference calculations in general require a significant increase in computer resources and calculation time. It is also hampered by the variability of selecting among the enormous number of configurations that quickly become available as the size of a system grows. As a result, single determinant ground state representations are by far more commonly performed. It should also be noted the two-electron repulsion terms are not summarily ignored as has been done in these examples, but are replaced by an effective one-electron mean field repulsion: $\hat{v}_i^{avg}$ as described for many-electron atoms in section 11.5. The molecular orbital wavefunctions are then numerically solved using the Hartree–Fock self-consistent field method. When the problem of interest requires a multi-reference ground state representation, there is an additional weighting coefficient for each determinant, which are then optimized in a variational procedure known as a *multi-configurational self-consistent field* (MCSCF) calculation.

There remains another hurdle inherent to numerical molecular calculations. When making a choice for an atomic orbital representation to expand either a molecular orbital or valence bond wavefunction, the hydrogenics (table 11.1) or Slater type orbitals (equation 11.33) are not viable forms for analytic determination of two-center integrals. Most commonly, basis functions are represented by Gaussian type orbitals (GTO). Gaussian functions have distinct computational advantages with regard to multi-center integral evaluation. This stems from the fact that Gaussians centered at two different points in space have a product which is a Gaussian located at a third point. The rnormalized 1s GTO is expressed in atomic units as:

$$\varphi = \left(\frac{2\alpha}{\pi}\right)^{3/4} e^{-\alpha r^2} \tag{13.39}$$

Those representing higher radial or angular momentum quantum numbers are multiplied by polynomials in r as well as spherical harmonic functions (see table 8.1). Coefficient α can be optimized by maximizing the spherical shell overlap between the 1s GTO and 1s ($\zeta = 1$) Slater type orbital (STO):

$$S = 4\pi \left(\frac{1}{\pi}\right)^{1/2} \left(\frac{2\alpha}{\pi}\right)^{3/4} \int_0^\infty r^2 e^{-r} e^{-\alpha r^2} \, dr \tag{13.40}$$

PARALLEL INVESTIGATION: Verify that the overlap between an STO and GTO is maximized with the coefficient: $\alpha = 0.271$.

Figure 13.5 shows plots of the amplitudes and probability densities of the $\zeta = 1$ STO and GTO using: $\alpha = 0.271$. It is immediately apparent the GTO is deficient in representing amplitude near the nucleus. It is said that GTOs lack the proper *cusp condition*. To remedy this shortcoming, approximate Slater functions (designated an STO-nG) are represented by linear combinations of Gaussians (known as *primitives*). The weighting factors, or *contraction coefficients*, are found in a least-squares fit rather than a variational procedure. Figure 13.6 compares plots of the optimized single Gaussian (STO-1G) with a two and three Gaussian fit of the form:

$$\text{STO–2G} = \left(\frac{2}{\pi}\right)^{3/4} [0.679(0.152)^{3/4}e^{-0.152r^2} + 0.430(0.852)^{3/4}e^{-0.852r^2}]$$

$$\text{STO–3G} = \left(\frac{2}{\pi}\right)^{3/4} \left[\begin{array}{l} 0.446(0.110)^{3/4}e^{-0.110r^2} + 0.535(0.406)^{3/4}e^{-0.406r^2} \\ 0.154(2.23)^{3/4}e^{-2.23r^2} \end{array}\right] \qquad (13.41)$$

A wide variety of more sophisticated options are available in quantum chemistry computational packages. Basis set choice and design is both an art and a science. Some terms commonly employed include: **Minimal basis**—use of one basis function representing an atomic orbital per electron for each atom comprising a molecular system. **Double-, triple-,... Zeta**—use of two, three, ... basis functions per atomic orbital. **Split valence**—use of one basis function per core orbital and multiple per valence orbital.

13.4 Qualitative molecular orbital theory for homonuclear diatomics

The molecular orbital picture uses atomic orbitals local to each nucleus to construct delocalized orbitals spanning the nuclear framework. Frequently, chemists employ a qualitative molecular orbital approach reminiscent of the model used for electron configurations in atoms. To designate wavefunctions, the Roman lettering system used for atoms is replaced by the appropriate Greek letter. It is based on the component of orbital angular momentum along the bond axis of the homonuclear diatomic: H_2^+, which if designated the z-Cartesian direction can be shown to have allowed values: $m_\ell = 0, \pm 1, \pm 2, ...$ One-electron molecular orbitals are then given the following assignments:

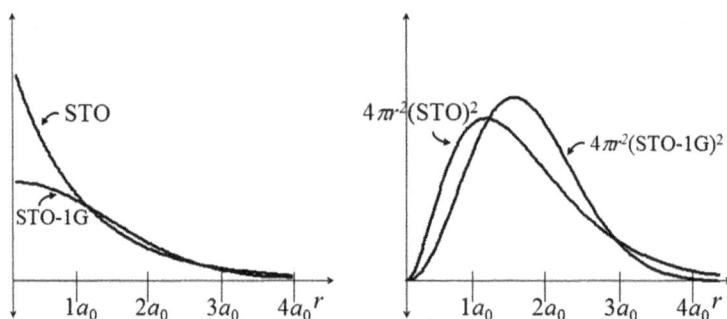

Figure 13.5. Comparison of STO and STO-1G amplitude and probability density shell.

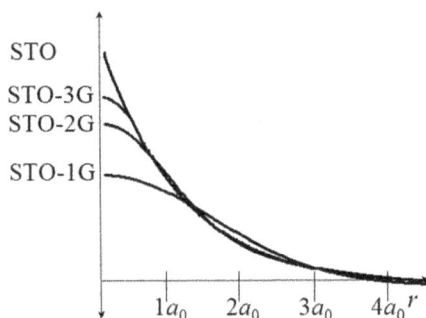

Figure 13.6. Comparison of STO, STO-1G, STO-2G, and STO-3G amplitudes.

$|m_\ell|$	0	1	2	3	$\ldots$
Letter	σ	π	δ	ϕ	$\ldots$

(13.42)

Furthermore, it is assumed that tightly-bound core orbitals remain localized on a nuclear center and do not become involved in molecular orbital formation. As a consequence, only valence electrons and orbitals are considered in the following description.

Let us begin with figure 13.7, the simple example of a homonuclear diatomic constructed from interaction between two 1s hydrogenics. According to equation (13.28), the basis functions superimpose with constructive or destructive interference, forming a bonding orbital now symbolized: σ_g and antibonding orbital denoted: σ_u. Orbital shading indicates the bonding molecular orbital is node-less over the nuclear framework, including the bonding region between centers. The difference in colors for the antibonding orbital indicates a change in phase of wavefunction amplitude, and includes a nodal plane in the region between centers. The orbital subscripts represent their parity relative to inversion through the bond center point. That assigned 'g' (German for *gerade*) is even, or does not change sign upon inversion about this point while 'u' (German for *ungerade*) is odd, showing a change in amplitude upon inversion.

Molecular orbital occupation is done in analogous fashion to the description of section 11.4. According to the Pauli exclusion principle, the fermion character of electrons permits a pair to occupy each molecular orbital so long as their spins oppose. This along with the *aufbau prinzip*, and Hund's rule (where required) determine occupation in a molecular configuration. For example, the ground state for H_2 is achieved by representing its two electons in one of the following expressions, depending on your taste: σ_g^2, $1\sigma_g^2$, or: $1s\sigma_g^2$. To further aid interpretation, the *bond order* (b.o.) is defined:

$$\text{b.o.} = \frac{1}{2} \, (\text{\#bonding } e^- \text{---\#antibonding } e^-) \tag{13.43}$$

The hydrogen molecule for instance has: b.o. $= \frac{1}{2} \, (2-0) = 1$ or is singly-bonded.

Molecular orbital designations for the isoelectronic cases: He_2^+ or H_2^- are $\sigma_g^2 \sigma_u^1$. A problem is encountered for either the hydrogen molecule dianion or neutral diatomic helium. In either case, the bond order is: $\frac{1}{2} \, (2-2) = 0$, hence molecular orbital theory

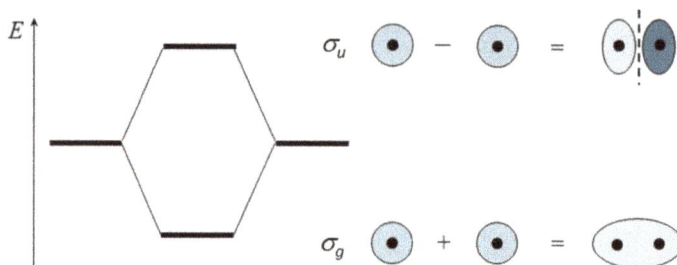

Figure 13.7. Formation of homonuclear $s\sigma$ orbitals.

claims neither exists. This can be traced back to equation (13.21) and the ensuing discussion explaining that orbitals of antibonding character are elevated above the separated atoms to a greater extent than bonding orbitals are depressed. Indeed, there is no experimental evidence of these species.

Moving to $\ell \geqslant 1$ orbitals, let us initially assume there is a sufficient energy gap such that there is no interaction between orbitals with different ℓ values. However, there are a multitude of possibilities for interactions of the p–p, d–d, and f–f type between centers. Determining the coincidence of a p orbital centered on nucleus A to one on B requires evaluating the overlap integral: $S_{AB} = \int d\tau p_A p_B$, but a qualitative graphical argument can be done in its place. Consider the two-center overlap of a p_z centered on nucleus A with a p_y on nucleus B, depicted in figure 13.8. Integration involves the infinite sum of product amplitude of the two functions. As figure 13.8 shows, the two orbitals exhibit two equivalent volumes of oppositely-signed product amplitude, which when summed over all space amounts to zero overlap. A similar argument can be made for homonuclear two center p_z–p_x and p_x–p_y overlap, as well that between d or f orbitals of differing Cartesian symmetry.

With the bond axis between centers taken along the z-direction, it is then apparent that p_z functions exhibit the greatest overlap, since they will have lobes in greatest proximity. As figure 13.9 demonstrates these orbitals interact strongest, producing a bonding σ_g and antibonding σ_u. The side-to-side orientation of the p_x and p_y allow two regions of either constructive or destructive interference depending on amplitude phase. These interaction regions appear both above and below the z-axis with either bonding or antibonding character. The resulting molecular orbital is designated as: π_u or π_g, respectively. The two π_u orbitals are lowered in energy by equal amounts due to the enhanced electron density created between centers by p-orbital overlap. In contrast, two degenerate π_g orbitals are formed by destructively-interfering p functions, creating a nodal plane perpendicular to the bond axis.

These orbitals are used to predict the valence molecular orbital configurations of second row homonuclear diatomics. Let us first consider the straightforward case of O_2, F_2, and Ne_2, which have a large energy gap between 2s and 2p that results from double occupation of orbitals from the p set. In this instance the orbital diagrams from figures 13.7 and 13.9 simply stack on top of one another as depicted in figure 13.10(a). Using the *aufbau prinzip* and Pauli exclusion principle, the valence electron configuration for O_2 is: $1\sigma_g^2 1\sigma_u^2 2\sigma_g^2 1\pi_u^4 1\pi_g^2$, with a bond order of: ½ (8–4) = 2.

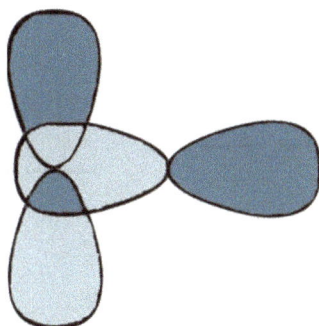

Figure 13.8. Graphical two-center overlap of a p_z and p_y orbital.

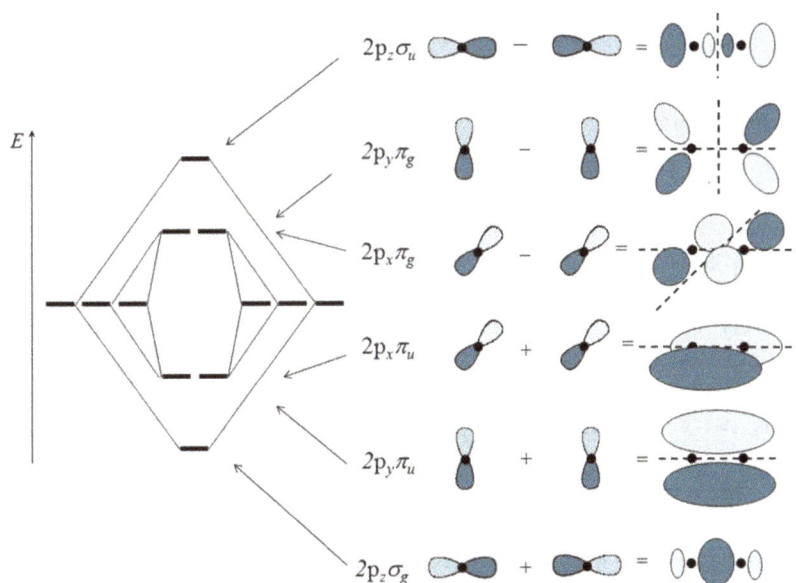

Figure 13.9. Formation of homonuclear $p\sigma$ and $p\pi$ orbitals.

The predicted double bond is a familiar feature of oxygen from Lewis dot representations in freshman chemistry.

PARALLEL INVESTIGATION: Verify that F_2 has a bond order = 1, or single bond.

Applying Hund's rule to the partially-occupied antibonding π_g further suggests oxygen has a triplet ground state. Oxygen is known to exhibit paramagnetic character, a verification of this spin state. Molecular orbital theory also predicts

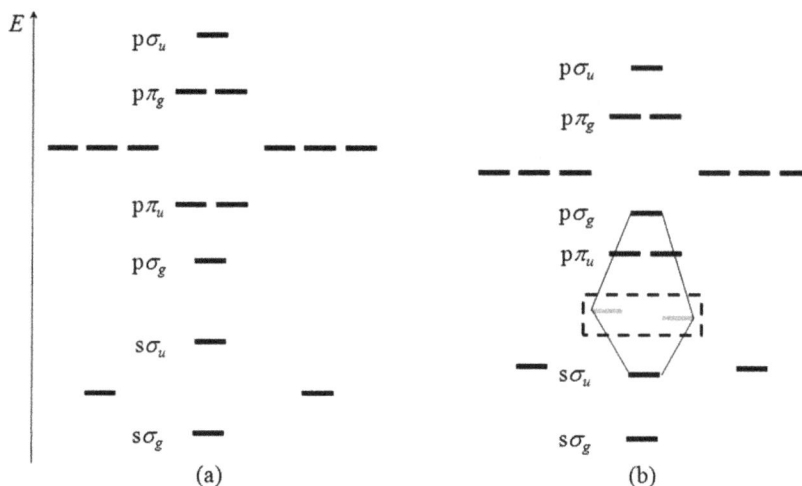

Figure 13.10. (a) Molecular orbital ordering for atoms with large s–p energy gap. (b) Molecular orbital ordering for atoms with small s–p energy gap.

that Ne_2 has zero bond order and therefore does not exist. There is in fact no evidence of this diatomic.

Molecular orbital energy diagrams for boron through nitrogen are complicated by a relatively small energy gap between 2s and 2p orbitals in these atoms. Since Hund's rule allows electrons to singly occupy degenerate p orbitals, the s–p gap is low enough to allow s–p interaction during molecular orbital formation. Assuming a z-bond axis, two center s and p_x or p_y overlap is zero for the same reasons described in figure 13.8. Symmetry does allow s and p_z interaction, however. The result of mixing, as shown in figure 13.10 (b), is an elevation of the $p\sigma_g$ to such an extent that it is now higher in energy than the π_u. For instance, the molecular orbital diagram for N_2 is: $1\sigma_g^2 1\sigma_u^2 1\pi_u^4 2\sigma_g^2$ with a predicted bond order of ½ (8−2) = 3, or triple bond.

The homonuclear diatomic model can be extended to transition metal d-type orbitals. In this case there are five functions interacting on each center, producing ten molecular orbitals. A bonding and antibonding σ of highest coincidence and thus maximum interaction results from d_{z^2} orbitals. The d_{xz} and d_{yz} form a degenerate bonding and antibonding set of π functions. The symmetry of the d_{xy} and $d_{x^2-y^2}$ orbitals produces a degenerate set of higher angular momentum molecular orbitals: the bonding δ_g and antibonding δ_u. Figure 13.11 shows examples of d orbital interactions and molecular orbital energetics. When forming a σ-type function, atomic orbitals on different center overlap in one region of space. The π interaction is characterized by two, and δ four distinct volumes of atomic orbital overlap. The pattern of adding molecular orbitals with sequentially increasing orbital angular momentum according to equation (13.42) continues upon mixing atomic orbitals of higher ℓ value. For instance, f-orbital interaction produces a bonding and anti-bonding σ, degenerate π, degenerate δ, now along with a degenerate ϕ set of molecular orbitals.

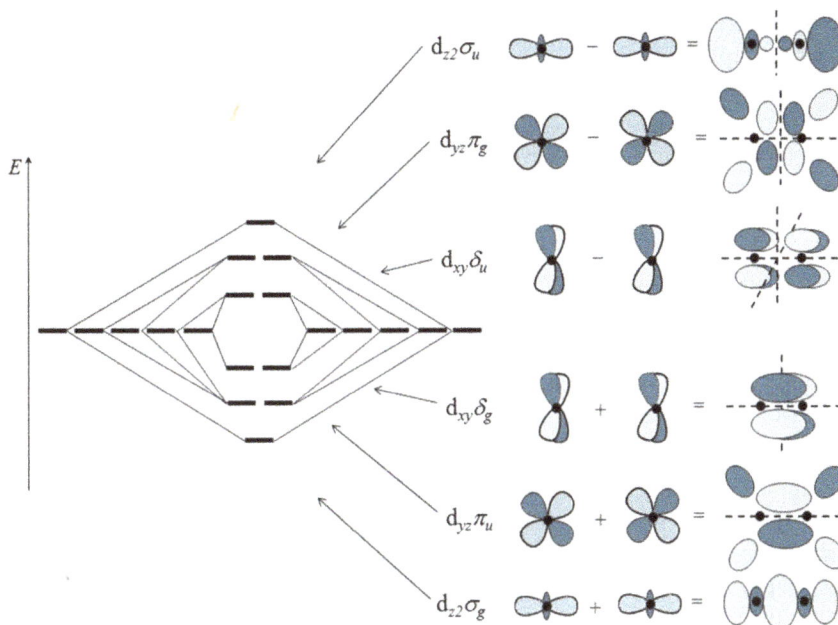

$d_{z^2}\sigma_u$

$d_{yz}\pi_g$

$d_{xy}\delta_u$

$d_{xy}\delta_g$

$d_{yz}\pi_u$

$d_{z^2}\sigma_g$

E

Figure 13.11. Formation of homonuclear dσ, dπ, and dδ orbitals.

The total spatial and spin angular momentum is combined in a molecular term symbol in a similar fashion to that which is done for atoms. Multi-electron spin degeneracy is handled in the same way as described in section 11.7. The total molecular orbital angular momentum is:

$$M_\ell = \sum_i m_{\ell_i} \qquad (13.44)$$

where the individual molecular orbital angular momenta are given in equation (13.42). Non-degenerate σ orbitals contribute no orbital angular momentum to the state, and, as is the case for atoms, a filled molecular orbital has cancelling orientations of the z-component, and no net orbital angular momentum as well. Once equation (13.44) is determined for each occupied molecular orbital, a capital Greek letter designates the total multi-electron spatial angular momentum. A left superscript gives the spin multiplicity, and a right subscript is used to represent the overall product of inversion symmetry for all electrons. For instance, H_2 has: $^1\Sigma_g$ ground states, while a configuration such as: $1\sigma_g^2 1\sigma_u^2 2\sigma_g^2 1\pi_u^1$ would be: $^2\Pi_u$.

13.5 The Hückel method

A simplified molecular orbital approach was proposed to describe the delocalized π systems of conjugated chain or aromatic ring molecules. This scheme involves a truncated variational model problem, using a Hamiltonian that dramatically simplifies the size of the matrix problem. A single Hamiltonian matrix element describes the contribution of each π molecular orbital in the system. In addition, there is a complete neglect of orbital overlap. Instead of analytically evaluating matrix elements over the molecular Hamiltonian, diagonal elements are parametrically assigned entries: $\alpha-E$, reminiscent of the self-interaction integral from one-electron molecular orbital theory (see equation (13.9)). In the same vein, nearest neighbor interactions between bonded atoms in the extended π system are expressed as parametric resonance integrals symbolized by: β (see equation (13.10)). All α values are equivalent, to each other, with the same approximation for β values. Members of the extended π system that are not directly bonded have Hamiltonian matrix elements set equal to zero.

The simplest possible Hückel molecule is ethene: CH_2CH_2. The two carbon atoms each contribute one orbital to the π system, giving a 2×2 Hamiltonian matrix. The solution to this problem parallels the case for the hydrogen molecule ion in equations (13.22)–(13.28). The ground state energy is determined by doubly occupying two π electrons in the lowest energy ethene orbital: $E_{tot} = 2(\alpha + \beta)$. This quantity is used as a reference point to measure the degree of resonance stabilization provided by a conjugated organic system. The ethene ground state Hückel energy can be envisioned as the energy of an isolated π bond in an organic molecule, thus providing a scale to determine the delocalization energy ε of an arbitrary N electron π system. The recipe for this is:

$$\varepsilon = E_{tot} - N(\alpha + \beta) \tag{13.45}$$

Using for example four π electron in 1,3-butadiene, we obtain the Hamiltonian matrix:

$$\mathbf{H} = \begin{vmatrix} \alpha - E & \beta & 0 & 0 \\ \beta & \alpha - E & \beta & 0 \\ 0 & \beta & \alpha - E & \beta \\ 0 & 0 & \beta & \alpha - E \end{vmatrix} \tag{13.46}$$

Following the development of equations (13.23)–(13.26), we arrive at the simplified Hamiltonian matrix:

$$\mathbf{H'} = \begin{vmatrix} 0 & 1 & 0 & 0 \\ 1 & 0 & 1 & 0 \\ 0 & 1 & 0 & 1 \\ 0 & 0 & 1 & 0 \end{vmatrix} \tag{13.47}$$

Which is diagonalized by eigenvector matrix:

$$\mathbf{C} = \begin{vmatrix} 0.372 & 0.602 & 0.602 & -0.372 \\ 0.602 & 0.372 & -0.372 & 0.602 \\ 0.602 & -0.372 & -0.372 & -0.602 \\ 0.372 & -0.602 & 0.602 & 0.372 \end{vmatrix} \tag{13.48}$$

with eigenvalues of the matrix:

$$\mathbf{X} = \begin{vmatrix} 1.618 \\ 0.618 \\ -0.618 \\ -1.618 \end{vmatrix} \qquad (13.49)$$

When equation (13.49) is used with equation (13.23) the π orbital energies are: $E_1 = \alpha + 1.618\beta$, $E_2 = \alpha + 0.618\beta$, $E_3 = \alpha - 0.618\beta$, and $E_4 = \alpha - 1.618\beta$. Doubly-occupying the two lowest energy orbitals with four π electrons gives total energy: $E_{tot} = 4\alpha + 4.472\beta$, and using equation (13.45) we obtain delocalization energy: $\varepsilon = 0.472\beta$. Column elements of the orthogonal transformation matrix given by equation (13.48) are used to expand the four orthonormal π molecular orbitals. These are qualitatively depicted with orbital phase only in figure 13.12.

PARALLEL INVESTIGATION: Verify the following Hückel delocalization energies for linear conjugated organic molecules:

System	1,3,5-hexatriene	1,3,5,7-octatetrene	1,3,5,7,9-decapentene
ε	0.988β	1.516β	2.056β

As an example of an organic ring system, consider the Hückel Hamiltonian matrix of cyclobutadiene. It is similar in appearance to 1,3-butadiene, differing by only two additional resonance integrals:

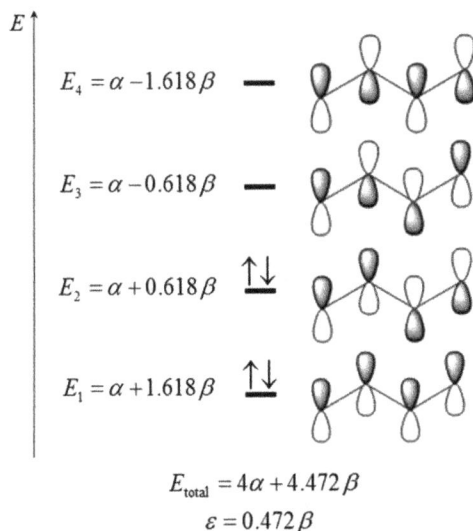

$$E_{total} = 4\alpha + 4.472\beta$$
$$\varepsilon = 0.472\beta$$

Figure 13.12. Hückel π molecular orbital system for 1,3-butadiene.

$$\mathbf{H} = \begin{vmatrix} \alpha - E & \beta & 0 & \beta \\ \beta & \alpha - E & \beta & 0 \\ 0 & \beta & \alpha - E & \beta \\ \beta & 0 & \beta & \alpha - E \end{vmatrix} \tag{13.50}$$

This system has eigenvector matrix:

$$\mathbf{C} = \begin{vmatrix} 0.500 & 0.707 & 0 & -0.500 \\ 0.500 & 0 & -0.707 & 0.500 \\ 0.500 & -0.707 & 0 & -0.500 \\ 0.500 & 0 & 0.707 & 0.500 \end{vmatrix} \tag{13.51}$$

which diagonalizes $\mathbf{H}'$ to give the matrix:

$$\mathbf{X} = \begin{vmatrix} 2 \\ 0 \\ 0 \\ -2 \end{vmatrix} \tag{13.52}$$

When equation (13.49) is used with equation (13.23) the π orbital energies are: $E_1 = \alpha + 2.00\beta$, $E_2 = \alpha$, $E_3 = \alpha$, and $E_4 = \alpha - 2.00\beta$. Doubly-occupying the two lowest energy orbitals with four π electrons gives total energy: $E_{\text{tot}} = 4\alpha + 4\beta$, and using equation (13.45) we obtain delocalization energy: $\varepsilon = 0$. The ring strain of this ring system does not permit the correct orbital positioning to facilitate aromatic character.

PARALLEL INVESTIGATION: Verify the Hückel delocalization energy for benzene is: $\varepsilon = 2.000\beta$ and naphthalene is: $\varepsilon = 3.684\beta$

It is common practice to designate π molecular orbitals based on their symmetry classification, using lower case point group irreducible representation labels based on the transformation properties of that orbital. Identifying the D_{4h} point group for cyclobutadiene (see table 13.1), the four p-orbitals used in π formation can be simultaneously manipulated according to operations of the group, with the sum of the *character* of these operations forming a *reducible representation*:

D_{4h}	E	$2C_4$	C_2	$2C_2'$	$2C_2''$	i	$2S_4$	σ_h	$2\sigma_v$	$2\sigma_d$
	4	0	0	0	−2	0	0	−4	0	2

Projection operator techniques are used to reduce this representation and find the four π molecular orbital *symmetry-adapted linear combinations*. These are: a_{2u} b_{1u}, and the doubly-degenerate e_g. This reduction can be easily verified by referring to table 13.1 to find the sum of these three irreducible representations. Using coefficients of column vectors in equation (13.51) to represent the phase of p-orbital contribution to a particular π orbital gives pictorial representations shown in figure 13.13. Appropriate manipulation and comparison with table 13.1 confirms the orbital with energy E_1 has symmetry a_{2u}, and that with E_4 transforms as b_{1u}.

To verify transformation properties of degenerate orbitals corresponding to E_2 and E_3, both must be simultaneously considered in the E_g irreducible representation. To see how this works, we find 2×2 matrices that transform each orbital under operations in the D_{4h} point group. Referring to orbitals labeled in figure 13.12, we find the following behavior for their simultaneous manipulation under each operation, and the 2×2 matrix which achieves this:

$$\begin{vmatrix} A \\ B \end{vmatrix} E \to \begin{vmatrix} A \\ B \end{vmatrix}\begin{vmatrix} A \\ B \end{vmatrix} \times \begin{vmatrix} 1 & 0 \\ 0 & 1 \end{vmatrix} = \begin{vmatrix} A \\ B \end{vmatrix}\begin{vmatrix} A \\ B \end{vmatrix} \quad C_4 \to \begin{vmatrix} B \\ A \end{vmatrix}\begin{vmatrix} A \\ B \end{vmatrix} \times \begin{vmatrix} 0 & 1 \\ 1 & 0 \end{vmatrix} = \begin{vmatrix} B \\ A \end{vmatrix}$$

$$\begin{vmatrix} A \\ B \end{vmatrix} C_2 \to \begin{vmatrix} -A \\ -B \end{vmatrix}\begin{vmatrix} A \\ B \end{vmatrix} \times \begin{vmatrix} -1 & 0 \\ 0 & -1 \end{vmatrix} = \begin{vmatrix} -A \\ -B \end{vmatrix}\begin{vmatrix} A \\ B \end{vmatrix} \quad C'_2 \to \begin{vmatrix} -B \\ -A \end{vmatrix}\begin{vmatrix} A \\ B \end{vmatrix} \times \begin{vmatrix} 0 & -1 \\ -1 & 0 \end{vmatrix} = \begin{vmatrix} -B \\ -A \end{vmatrix}$$

$$\begin{vmatrix} A \\ B \end{vmatrix} C''_2 \to \begin{vmatrix} -A \\ B \end{vmatrix}\begin{vmatrix} A \\ B \end{vmatrix} \times \begin{vmatrix} -1 & 0 \\ 0 & 1 \end{vmatrix} = \begin{vmatrix} -A \\ B \end{vmatrix}\begin{vmatrix} A \\ B \end{vmatrix} \quad i \to \begin{vmatrix} A \\ B \end{vmatrix}\begin{vmatrix} A \\ B \end{vmatrix} \times \begin{vmatrix} 1 & 0 \\ 0 & 1 \end{vmatrix} = \begin{vmatrix} A \\ B \end{vmatrix}$$

$$\begin{vmatrix} A \\ B \end{vmatrix} S_4 \to \begin{vmatrix} -B \\ A \end{vmatrix}\begin{vmatrix} A \\ B \end{vmatrix} \times \begin{vmatrix} 0 & -1 \\ 1 & 0 \end{vmatrix} = \begin{vmatrix} -B \\ A \end{vmatrix}\begin{vmatrix} A \\ B \end{vmatrix} \quad \sigma_h \to \begin{vmatrix} -A \\ -B \end{vmatrix}\begin{vmatrix} A \\ B \end{vmatrix} \times \begin{vmatrix} -1 & 0 \\ 0 & -1 \end{vmatrix}\begin{vmatrix} -A \\ -B \end{vmatrix}$$

$$\begin{vmatrix} A \\ B \end{vmatrix} \sigma_v \to \begin{vmatrix} B \\ A \end{vmatrix}\begin{vmatrix} A \\ B \end{vmatrix} \times \begin{vmatrix} 0 & 1 \\ 1 & 0 \end{vmatrix} = \begin{vmatrix} B \\ A \end{vmatrix}\begin{vmatrix} A \\ B \end{vmatrix} \quad \sigma_d \to \begin{vmatrix} A \\ -B \end{vmatrix}\begin{vmatrix} A \\ B \end{vmatrix} \times \begin{vmatrix} 1 & 0 \\ 0 & -1 \end{vmatrix}\begin{vmatrix} A \\ -B \end{vmatrix}$$

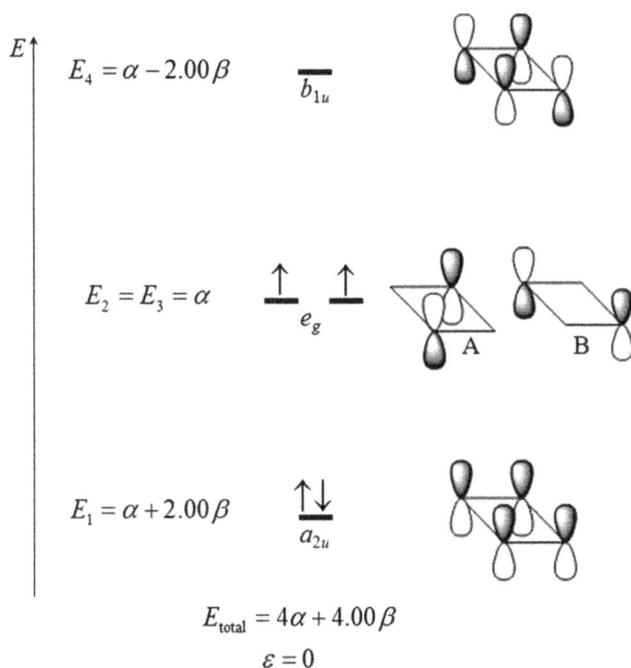

$E_4 = \alpha - 2.00\beta \qquad b_{1u}$

$E_2 = E_3 = \alpha \qquad e_g \qquad$ A $\qquad$ B

$E_1 = \alpha + 2.00\beta \qquad a_{2u}$

$E_{total} = 4\alpha + 4.00\beta$

$\varepsilon = 0$

Figure 13.13. Hückel π molecular orbital system for cyclobutadiene.

Table 13.1. The D_{4h} point group.

D_{4h}	E	$2C_4$	C_2	$2C_2'$	$2C_2''$	i	$2S_4$	σ_h	$2\sigma_v$	$2\sigma_d$	
A_{1g}	1	1	1	1	1	1	1	1	1	1	x^2+y^2, z^2
A_{2g}	1	1	1	−1	−1	1	1	1	−1	−1	R_z
B_{1g}	1	−1	1	1	−1	1	−1	1	1	−1	x^2-y^2
B_{2g}	1	−1	1	−1	1	1	−1	1	−1	1	xy
E_g	2	0	−2	0	0	2	0	−2	0	0	(R_x, R_y) (xz, yz)
A_{1u}	1	1	1	1	1	−1	−1	−1	−1	−1	
A_{2u}	1	1	1	−1	−1	−1	−1	−1	1	1	z
B_{1u}	1	−1	1	1	−1	−1	1	−1	−1	1	
B_{2u}	1	−1	1	−1	1	−1	1	−1	1	−1	
E_u	2	0	−2	0	0	−2	0	2	0	0	(x, y)

Characters of the E_g irreducible representation are found by taking the trace of each of the above transformation matrices:

D_{4h}	E	$2C_4$	C_2	$2C_2'$	$2C_2''$	i	$2S_4$	σ_h	$2\sigma_v$	$2\sigma_d$
E_g	2	0	−2	0	0	2	0	−2	0	0

Hückel molecular orbital theory can also be applied to fullerene molecules such as: C_{20}. The molecule, shown in figure 13.14, possesses icosahedral symmetry. A p-orbital on each carbon contributes to a delocalized π system. The Hückel approach is applied following equations (13.22)–(13.26). Rather than show the 20 × 20 Hamiltonian or coefficient matrix, we cut right to results as depicted in figure 13.15. The lowest energy eigenvalue represents a non-degenerate orbital of a_g symmetry with energy $\alpha + 3\beta$. This is followed by a sequence of orbitals with high degeneracy. Eigenvalue E_2 is triply-degenerate, describing a t_{1u} orbital with energies: $\alpha + 2.24\beta$. Next is a quintuple-degenerate h_g orbital of energy: $\alpha + \beta$. These orbitals are shown in figure 13.16.

Before populating the orbitals of C_{20}, we give mention to Hückel's rule of aromaticity. This states that monocyclic ring systems possessing 4N +2 pi electrons exhibit unique stability and reactivity, and are collectively known as *aromatic* molecules. Among these properties are equivalent C–C bond lengths for all ring members, and a high resistance to electrophilic addition reactions in which double bonds are broken and replaced by saturated single bonds.

The 20 π electrons of C_{20} do not satisfy the 4N +2 rule. However, the C_{20}^{+2} dication does with $N = 4$, as well as C_{20}^{-2} dianion, with $N = 5$. When equilibrium geometry calculations are performed, three distinct bond lengths are found for the neutral species. Similar calculation for the C_{20} dication however produce the same C–C bond length for all pairs, whether a sophisticated or relatively low-cost calculation is performed. This is encouraging, given that it is well known that aromatic bonds are equivalent for all carbon atoms. Unfortunately, the C_{20}

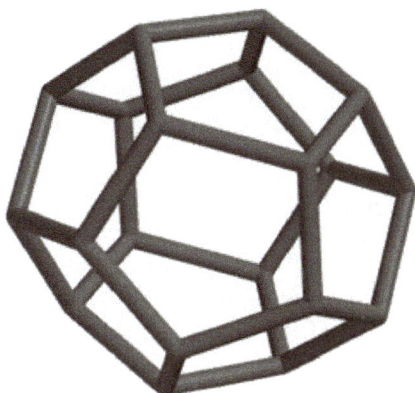

Figure 13.14. The icosahedral fullerene: C_{20}.

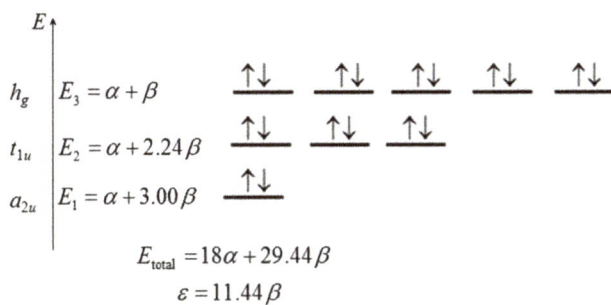

$$E_{total} = 18\alpha + 29.44\beta$$
$$\varepsilon = 11.44\beta$$

Figure 13.15. Hückel π molecular orbital system for C_{20}^{+2}.

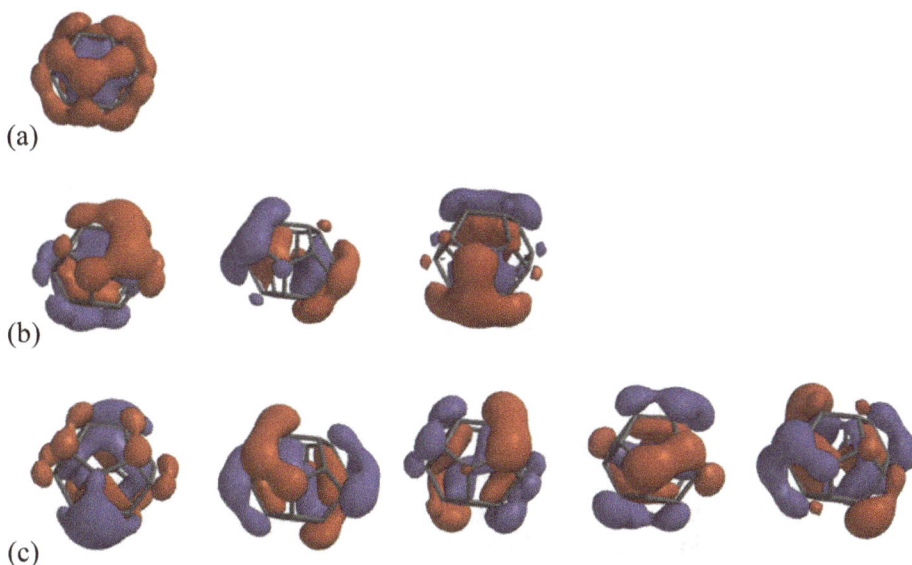

(a)

(b)

(c)

Figure 13.16. (a) The C_{20} a_g orbital. (b). The C_{20} t_{1u} orbitals. (c) The C_{20} g_u orbitals.

13-23

dianion again gives three distinct bond lengths, in contrast to predictions of the $4N + 2$ rule.

As a possible alternative interpretion of Hückel's rule, $4N + 2$ π electrons is also an appropriate occupation for energy levels of a particle on a ring model as described in section 8.1.

$$E = \frac{\hbar^2 m_\ell^2}{2I} \qquad m_\ell = 0, \pm1, \pm2,... \tag{13.53}$$

Based on allowed values of m_ℓ, these orbitals have the same degeneracy as do the Hückel rule. It is not a stretch of the imagination to envision this type of behavior for electrons populating the delocalized π molecular orbitals of an aromatic ring system. A particle on a ring model is not applicable to three-dimensional fullerenes such as C_{20}, however its spatial symmetry has similarities to the particle on a sphere discussed in section 8.2. These solutions have energy levels and degeneracies:

$$E = \frac{\ell(\ell + 1)\hbar^2}{2I} \quad \ell = 0, 1, 2, ... \tag{13.54}$$

Recalling from section 8.2, particle on a sphere energy levels of a given ℓ value have degeneracy according to: $m_\ell = 0, \pm1, \pm2, ... \pm\ell$. The first is thus non-degenerate, the second triply degenerate, and third quintuple degenerate (see figure 8.2), which is the same degeneracy structure the C_{20} Hückel energy levels exhibit in figure 13.15.

Considering the π electrons of neutral, dianion, and dication forms of C_{20}, it is noted that only the dication has the appropriate number to completely occupy either the first three particle on a sphere energy levels. When the C_{20} orbitals are populated with the 18 electrons of the dication, we find a total energy of: $18\alpha + 29.44\beta$ and a very large delocalization energy of 11.44β. It was previously noted the $C_{20}{}^{+2}$ species also has all equal bond lengths as is true of monocyclic aromatics, but is this a mere coincidence? More evidence is supplied by computing electrostatic potential maps: V_p for the three C_{20} systems. This quantity is defined as the energy of a positive point charge located in space at point: p as it interacts with the nuclei and electrons of a molecule. This quantity therefore takes the form of a Coulombic potential term over the nuclei and molecular orbitals ϕ_i:

$$V_p = \sum_a^{nuclei} \frac{Z_A e^2}{4\pi\varepsilon_0 R_{Ap}} - \frac{e^2}{4\pi\varepsilon_0} \sum_{i,j} \int d\tau \frac{\varphi_i^* \varphi_j}{r} \tag{13.55}$$

Figure 13.17 compares electrostatic potential maps for the two cyclic hydrocarbon hexagonal ring systems: benzene and 1,3-cyclohexadiene. Cyclohexadiene differs from benzene by one of the three possible double bonding sites being saturated by the addition of H_2. The electrostatic potential map for these two shows a distinctly even charge distribution predicted for the aromatic system. When similar charge

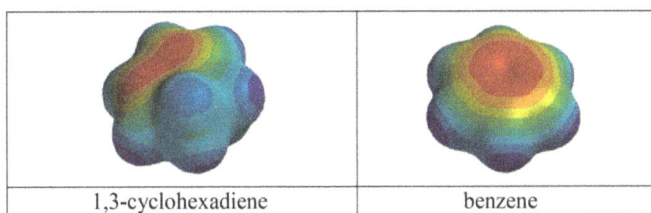

Figure 13.17. Electrostatic potential maps for 1,3,cyclohexadiene (C_6H_8) and benzene (C_6H_6).

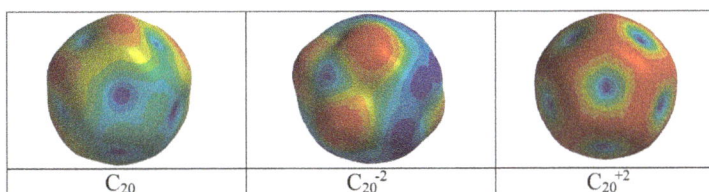

Figure 13.18. Electrostatic potential maps of C_{20} systems.

distribution maps are prepared for C_{20} along with its dianion and dication, it is evident that only the positively charged species, the one which has the π electron system satisfying a particle on a sphere criterion, also exhibits electrostatic potential of symmetric appearance.

IOP Concise Physics

What's the Matter with Waves?
An introduction to techniques and applications of quantum mechanics
William Parkinson

Appendix A

Physical constants and units

A.1 Atomic units

In quantum mechanical applications, physicists frequently simplify things by setting four of the most commonly employed physical constants each to values of exactly 1: (a) elementary unit of charge e, (b) rest mass of electron m_e, (c) vacuum permittivity $4\pi\varepsilon_0$, and (d) Planck's constant in the form angular momentum per cycle $\hbar = h/2\pi$. Furthermore, in the system known as *Hartree units*, the nucleus is taken to be infinitely massive compared to the electron so that according to equation (2.12): $\mu = m_e = 1$. In Hartree units the Bohr radius and Hartree energy unit therefore have values $a_0 = 1$ au and $E_H = 1$ au. (The unit 'au' or 'a.u.' stands for atomic units, not to be confused with *astronomical units*, for which the same symbol is employed. The irony should not be lost, representing the smallest and largest of dimensions with the same abbreviation.) Therefore, from equations (4.27) and (4.31) of chapter 4 we find the proton–electron inter-atomic distance in the hydrogen atom is: $r = a_0 = 1$ au and its ground state ($N = 1$) energy is -0.5 au. These equations also naturally define au to SI conversion factors for length (1 au $= 5.29 \times 10^{-11}$ m) and energy (1 au $= 4.36 \times 10^{-18}$ J).

Values in au for other common physical constants can be derived. For instance, in Hartree atomic units, the time unit is defined from: $\hbar/E_h$, which gives an *SI* conversion of: 1 au $= 2.42 \times 10^{-17}$ s. In chapter 12, we discuss the atomic polarizability, which has SI units: $C^2 \cdot m^2/E_h$. Conversion of this quantity to SI units is accomplished as follows:

$$\frac{(1.60 \times 10^{-19}\,C)^2}{au} \cdot \frac{(5.29 \times 10^{-11}\,m)^2}{au} \cdot \frac{au}{(4.36 \times 10^{-18}\,J)} \qquad (A.1)$$

$$= 1.64 \times 10^{-41}\,C^2 \cdot m^2 \cdot J^{-1}$$

doi:10.1088/978-1-6817-4577-0ch14

Table A.1. Physical constants

Quantity	Symbol	Value
Planck constant	$\hbar = h/2\pi$	1.05457×10^{-34} J · s
	h	6.62608×10^{-34} J · s
Mass of electron	m_e	9.10939×10^{-31} kg
Mass of proton	m_p	1.67262×10^{-27} kg
Elementary charge	e	1.60218×10^{-19} C
Vacuum permittivity	$4\pi\varepsilon_0$	1.11265×10^{-10} J^{-1} C^2 m^{-1}
Pi	π	3.14159
Avogadro's number	N_A	6.02214×10^{23} mol^{-1}
Speed of light	c	2.99792×10^{8} m · s^{-1}

When converted to cgs units by multiplication with 8.99×10^{15}, the polarizability unit is: 1 au = 0.147 (10^{24} cm^3), which is equivalently: 1 au = 0.147 Å^3.

The proton mass can be found from the ratio: m_p/m_e, giving $m_p \approx 1836$ au. Assuming an infinitely-massive nucleus results in a reduced mass of m_e and hence 1 in atomic units for the hydrogen atom, it also simplifies calculations relative to the kinetic energy contributions of the nucleus to the total energy of an atom itself in electronic structure calculations. To prevent any confusion quantities such as the Rydberg constant are symbolized as: R_∞ (for infinitely-massive nuclei) when it is represented by its value in Hartree atomic units. In some applications, the Rydberg constant takes the form:

$$R_M = R_\infty \frac{1}{1 + m_e/M} \tag{A.2}$$

Where M is the total mass of all nucleons in the nucleus (the correction term should be recognized as a means to incorporate the correct two-body reduced mass formula). The magnitude of this correction term is largest in the hydrogen atom itself, for as the nuclear mass increases, $\mu \to m_e$. Using values from table A.1, the reduced mass of the hydrogen atom is:

$$\mu = \frac{m_e m_p}{m_e + m_p} = \frac{(9.10939 \times 10^{-31}\,\text{kg}) \cdot (1.67262 \times 10^{-27}\,\text{kg})}{(9.10939 \times 10^{-31}\,\text{kg}) + (1.67262 \times 10^{-27}\,\text{kg})} \tag{A.3}$$
$$= 9.10443 \times 10^{-31}\,\text{kg}$$

which differs from the electron mass in the fourth significant digit.

IOP Concise Physics

What's the Matter with Waves?
An introduction to techniques and applications of quantum mechanics
William Parkinson

Appendix B

Calculus and trigonometry essentials

B.1 Differentiation

$$\frac{\mathrm{d}}{\mathrm{d}x}x^n = n \cdot x^{n-1} \tag{B.1}$$

$$\frac{\mathrm{d}}{\mathrm{d}x}\ln(y) = \frac{1}{y} \cdot \frac{\mathrm{d}y}{\mathrm{d}x} \tag{B.2}$$

$$\frac{\mathrm{d}}{\mathrm{d}x}e^y = e^y \cdot \frac{\mathrm{d}y}{\mathrm{d}x} \tag{B.3}$$

$$\frac{\mathrm{d}}{\mathrm{d}x}\sin^n(y) = n \cdot \sin^{n-1}(y) \cdot \cos(y) \cdot \frac{\mathrm{d}y}{\mathrm{d}x} \tag{B.4}$$

$$\frac{\mathrm{d}}{\mathrm{d}x}\cos^n(y) = -n \cdot \cos^{n-1}(y) \cdot \sin(y) \cdot \frac{\mathrm{d}y}{\mathrm{d}x} \tag{B.5}$$

$$\frac{\mathrm{d}}{\mathrm{d}x}\sin^{-1}(y) = \frac{1}{\sqrt{1-y^2}} \cdot \frac{\mathrm{d}y}{\mathrm{d}x} \tag{B.6}$$

$$\frac{\mathrm{d}}{\mathrm{d}x}\cos^{-1}(y) = -\frac{1}{\sqrt{1-y^2}} \cdot \frac{\mathrm{d}y}{\mathrm{d}x} \tag{B.7}$$

$$\frac{\mathrm{d}}{\mathrm{d}x}\tan^{-1}(y) = \frac{1}{1+y^2} \cdot \frac{\mathrm{d}y}{\mathrm{d}x} \tag{B.8}$$

doi:10.1088/978-1-6817-4577-0ch15   B-1   © Morgan & Claypool Publishers 2017

$$\frac{d}{dx} f(y) \cdot g(y) = g(y) \cdot \frac{df(y)}{dx} + f(y) \cdot \frac{dg(y)}{dx} \tag{B.9}$$

B.2 Integration

Indefinite integrals

$$\int x^n \, dx = \frac{1}{n+1} \cdot x^{n+1} + C \tag{B.10}$$

$$\int \ln(x) dx = \frac{1}{x} + C \tag{B.11}$$

$$\int \sin^2(ax) dx = \frac{x}{2} - \frac{1}{4a} \cdot \sin(2ax) + C \tag{B.12}$$

$$\int \cos^2(ax) dx = \frac{x}{2} + \frac{1}{4a} \cdot \sin(2ax) + C \tag{B.13}$$

$$\int \sin(ax) \cdot \cos(ax) dx = \frac{1}{2a} \sin^2(ax) + C \tag{B.14}$$

$$\int \sin(ax) \cdot \sin(bx) dx$$
$$= \frac{(a-b) \cdot \sin((a+b)x) - (a+b) \cdot \sin((a-b)x)}{2(a^2 - b^2)} + C \tag{B.15}$$

$$\int \cos(ax) \cdot \cos(bx) dx$$
$$= \frac{(a+b) \cdot \sin((a-b)x) + (a-b) \cdot \sin((a+b)x)}{2(a^2 - b^2)} + C \tag{B.16}$$

$$\int \sin^3(ax) dx = -\frac{1}{3a} \cdot \cos(ax) \cdot (\sin^2(ax) + 2) + C \tag{B.17}$$

$$\int \sin^n(ax) \cdot \cos(ax) dx = \frac{\sin^{n+1}(ax)}{(n+1)a} + C \tag{B.18}$$

$$\int x \cdot \sin^2(ax) dx = \frac{x^2}{4} - \frac{x \cdot \sin(2ax)}{4a} - \frac{\cos(2ax)}{8a^2} + C \tag{B.19}$$

$$\int x \cdot \cos^2(ax)dx = \frac{x^2}{4} + \frac{x \cdot \sin(2ax)}{4a} + \frac{\cos(2ax)}{8a^2} + C \qquad \text{(B.20)}$$

$$\int x \cdot \sin(ax) \cdot \sin(bx)dx = \frac{1}{2}\left[\frac{x \cdot \sin((a-b)x)}{a-b} - \frac{x \cdot \sin((a+b)x)}{a+b}\right]$$
$$+ \frac{1}{2}\left[\frac{\cos((a-b)x)}{(a-b)^2} - \frac{\cos((a+b)x)}{(a+b)^2}\right] + C \qquad \text{(B.21)}$$

$$\int x \cdot \cos(ax) \cdot \cos(bx)dx = \frac{1}{2}\left[\frac{x \cdot \sin((a-b)x)}{a-b} + \frac{x \cdot \sin((a+b)x)}{a+b}\right]$$
$$+ \frac{1}{2}\left[\frac{\cos((a-b)x)}{(a-b)^2} + \frac{\cos((a+b)x)}{(a+b)^2}\right] + C \qquad \text{(B.22)}$$

$$\int x^2 \cdot \sin^2(ax)dx = \frac{x^3}{6} - \left(\frac{x^2}{4a} - \frac{1}{8a^3}\right)\sin(2ax) - \frac{x \cdot \cos(2ax)}{4a^2} + C \qquad \text{(B.23)}$$

$$\int x^2 \cdot \cos^2(ax)dx = \frac{x^3}{6} + \left(\frac{x^2}{4a} - \frac{1}{8a^3}\right)\sin(2ax) + \frac{x \cos(2ax)}{4a^2} + C \qquad \text{(B.24)}$$

$$\int (x-c)^2 \cdot \sin^2(ax)dx = \frac{4a^3x \cdot (3c^2 - 3cx + x^2)}{24a^3}$$
$$+ \frac{(3 - 6a^2 \cdot (c-x)^2) \cdot \sin(2ax) + 6a \cdot (c-x) \cdot \cos(2ax)}{24a^3} + C \qquad \text{(B.25)}$$

$$\int (x-c)^2 \cdot \sin(ax) \cdot \sin(bx)dx$$
$$= \frac{1}{2}\left[\frac{(a^2x^2 - 2abx^2 + b^2x^2 - 2) \cdot \sin((a-b)x)}{(a-b)^3}\right]$$
$$- \frac{1}{2}\left[\frac{(a^2x^2 + 2abx^2 + b^2x^2 - 2) \cdot \sin((a+b)x)}{(a+b)^3} - \frac{c^2 \sin((a-b)x)}{a-b}\right]$$
$$- \frac{1}{2}\left[\frac{2c \cdot ((a-b)x \cdot \sin((a-b)x) + \cos((a-b)x))}{(a-b)^2} + \frac{c^2 \sin((a+b)x)}{a+b}\right]$$
$$+ \frac{1}{2}\left[\frac{2c \cdot ((a+b)x \cdot \sin((a+b)x) + \cos((a+b)x))}{(a+b)^2}\right]$$
$$+ \frac{1}{2}\left[\frac{2x \cdot \cos((a-b)x)}{(a-b)^2} - \frac{2x \cdot \cos((a+b)x)}{(a+b)^2}\right] + C \qquad \text{(B.26)}$$

$$\int x^2 \cdot \cos(ax) \cdot \cos(bx)\mathrm{d}x$$

$$= \frac{1}{2}\left[\frac{(a^2x^2 - 2abx^2 + b^2x^2 - 2) \cdot \sin((a - b)x)}{(a - b)^3}\right]$$

$$+ \frac{1}{2}\left[\frac{(a^2x^2 + 2abx^2 + b^2x^2 - 2) \cdot \sin((a + b)x)}{(a + b)^3}\right]$$

$$+ \frac{1}{2}\left[\frac{2x \cdot \cos((a - b)x)}{(a - b)^2} + \frac{2x \cdot \cos((a + b)x)}{(a + b)^2}\right] + C$$

(B.27)

$$\int x \cdot \sin^3(ax)\mathrm{d}x = \frac{x \cdot \cos(3ax)}{12a} - \frac{\sin(3ax)}{36a^2}$$
$$- \frac{3x \cdot \cos(ax)}{4a} + \frac{3\sin(ax)}{4a^2} + C$$

(B.28)

Definite integrals

$$\int_0^\infty x^n \cdot e^{-ax}\,\mathrm{d}x = \frac{n!}{a^{n+1}} \qquad n\ \text{int}$$

(B.29)

$$\int_0^\infty e^{-ax^2}\,\mathrm{d}x = \frac{1}{2}\sqrt{\frac{\pi}{a}}$$

(B.30)

$$\int_0^\infty x \cdot e^{-ax^2}\,\mathrm{d}x = \frac{1}{2a}$$

(B.31)

$$\int_0^\infty x^2 \cdot e^{-ax^2}\,\mathrm{d}x = \frac{\sqrt{\pi}}{4a^{3/2}}$$

(B.32)

$$\int_0^\infty x^3 \cdot e^{-ax^2}\,\mathrm{d}x = \frac{1}{2a^2}$$

(B.33)

$$\int_0^\infty x^4 \cdot e^{-ax^2}\,\mathrm{d}x = \frac{3\sqrt{\pi}}{8a^{5/2}}$$

(B.34)

$$\int_0^\infty x^5 \cdot e^{-ax^2}\,\mathrm{d}x = \frac{1}{a^3}$$

(B.35)

$$\int_0^\infty x^6 \cdot e^{-ax^2}\,\mathrm{d}x = \frac{15\sqrt{\pi}}{16a^{7/2}}$$

(B.36)

$$\int_0^\infty x^{1/2} \cdot e^{-ax}\, dx = \frac{\sqrt{\pi}}{2a^{3/2}} \qquad\qquad (B.37)$$

$$\int_0^\infty x^{3/2} \cdot e^{-ax}\, dx = \frac{3\sqrt{\pi}}{4a^{5/2}} \qquad\qquad (B.38)$$

B.3 Trigonometric identities

$$e^{\pm i\theta} = \cos\theta \pm i\sin\theta \qquad\qquad (B.39)$$

$$\tan\theta = \frac{\sin\theta}{\cos\theta} \qquad\qquad (B.40)$$

$$\sin^2\theta + \cos^2\theta = 1 \qquad\qquad (B.41)$$

$$\sin^2\theta = \frac{1 - \cos 2\theta}{2} \qquad\qquad (B.42)$$

B.4 Functional symmetry and integration

A function which obeys the condition:

$$f(-x) = -f(x) \qquad\qquad (B.43)$$

is symmetric with respect to the origin of a Cartesian coordinate system and is known as an *odd function*. One which obeys the condition:

$$f(-x) = f(x) \qquad\qquad (B.44)$$

is symmetric with respect to the y-axis of a Cartesian coordinate system and is known as an *even function*.

Any polynomial containing only positive or negative odd powers of x is an odd function. Polynomials (or exponentials) with positive or negative even powers of x (including zero) are even functions. Plots of the important trigonometric functions show that sine is an odd function and cosine is even. Products of symmetric functions have symmetry which obey:

$$\text{even} \times \text{even} = \text{even} \quad \text{odd} \times \text{even} = \text{odd} \quad \text{odd} \times \text{odd} = \text{even} \qquad (B.45)$$

For instance, $\sin^2(x)$ is an even function, but $\cot(x) = \cos(x)/\sin(x)$ is odd. Examples of odd and even functions are presented in figures B.1 and B.2, respectively.

The symmetry of $f(x)$ plays a critical role in its definite integral over the range: $-a \leqslant x \leqslant +a$. The integration process involves infinite area summation. As is seen in figure B.2, an even function or product of functions has a mirror image form about

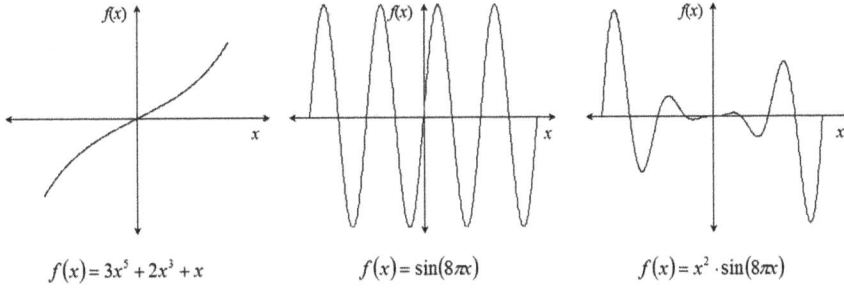

Figure B.1. Examples of odd functions.

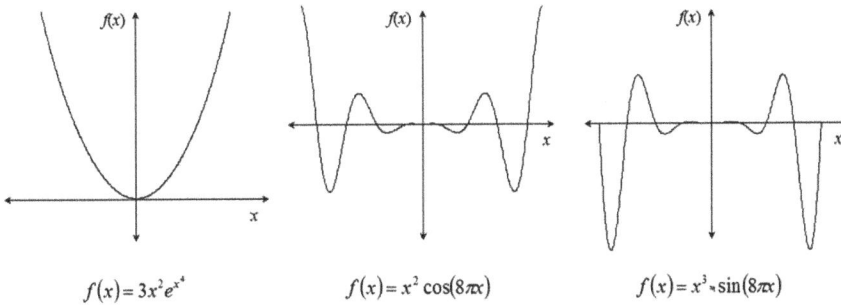

Figure B.2. Examples of even functions.

the y-axis. The area under the function's curve has the same phase on both sides of this axis. As a result, definite integrals of even functions obey the condition:

$$\int_{-a}^{+a} f(x)\,dx = 2 \cdot \int_{0}^{+a} f(x)dx \quad f(x) \text{ even} \tag{B.46}$$

As seen in figure B.1, odd functions or products which are overall odd have curves in opposite phase with respect to the y-axis. Because of their asymmetry, these functions have definite integrals which vanish over symmetric limits:

$$\int_{-a}^{+a} f(x)\,dx = 0 \quad f(x) \text{ odd} \tag{B.47}$$

These conclusions hold true for cases up to and including definite integration between positive and negative infinity.

Index

www.ingramcontent.com/pod-product-compliance
Lightning Source LLC
Chambersburg PA
CBHW082034230326
41598CB00081B/6346